To professor MIRON NICOLESCU,
on his 70th anniversary

ION SUCIU

FUNCTION ALGEBRAS

translated from the Romanian
by
Dr. Mihaela Mihăilescu

Editura Academiei Republicii Socialiste România București
1975
Noordhoff International Publishing Leyden

The monograph is a revised translation of the
original Romanian version
"Algebre de funcţii"
Editura Academiei, 1969

Softcover reprint of the hardcover 1st edition 1969

Work awarded the "Simion Stoilov" Prize of the Romanian Academy

ISBN-13: 978-90-286-0445-2 e-ISBN-13: 978-94-010-1912-5
DOI: 10.1007/978-94-010-1912-5

EDITURA ACADEMIEI REPUBLICII SOCIALISTE ROMÂNIA
Calea Victoriei 125, Bucureşti

Preface

Under the title of Function Algebras we may now include a very large number of works, published mainly in the last decade, which constitute one of the important chapters of functional analysis. This chapter has grown up from various problems, permanently furnished to mathematics, by the theory of functions, using modern methods of algebra, topology and functional analysis and presenting large possibilities of applications in operators theory. Herefrom proceeds its living character, the variety of obtained results, the variety of forms and contexts in which these results can be found. This also explains the difficulty of an exhaustive exposition of these problems.

The purpose of the monograph is to present a coherent exposition of the fundamental results of this theory with an orientation to their applicability to the theory of operator representations of function algebras. The idea of such a work appeared during the seminaries on function algebras held at the Mathematical Institute in Bucharest, under the direction of C. Foiaş and at the Faculty of Mathematics and Mechanics under the direction of N. Boboc.

It is a pleasure for the author to express his gratitude to C. Foiaş for assistance in his efforts, in general, and for the large contribution the discussions and cooperation with him had brought in the elaboration of this monograph.

I also would like to thank N. Boboc for the clear discussions we have had during the seminaries and the elaboration of some chapters. I am grateful to G. Gussi for the useful discussions on various problems and to all my seminary colleagues.

My thanks are due to the Publishing House of the Romanian Academy for the graphic presentation of the monograph.

I. S.

Bucharest, September 1968

Preface to the English edition

The English version of this monograph contains several changes and completions relative to the Romanian edition. I mention some of them: the proofs to the theorems on spectral and semispectral measures in paragraph 7.1, the Wold decomposition for subspectral representations (paragraph 8.4), the decompositions with respect to the Gleason parts (paragraph 8.5), the results relating Theorem 9.15 to the known predication theorems (paragraph 9.5). Chapter 10 was entirely rewritten in order to have a frame for the example of a function algebra representation with no spectral dilation.

I take this opportunity to thank professor B. Sz.-Nagy for the interest he took in my book as well as for his useful suggestions relative to completions.

In writing the completions I had continued to benefit from C. Foiaş advice. I thank him once more.

I also like to thank my colleague Mihaela Mihăilescu for translating this book into English.

I. S.

Contents

CHAPTER 1

Preliminaries

1.1. Commutative Banach algebras

Throughout this book we shall denote by C the field of complex numbers and by R the field of real numbers.

A *Banach algebra* over C is a non-empty set A endowed with a vector algebra structure over C and a norm relative to which the linear space A is a Banach space, with the property

$$\|ab\| \leqslant \|a\| \|b\| \qquad (a, b \in A). \tag{1.1.1}$$

Moreover, it is supposed that multiplication in A is commutative and that there exists a unit element $e \in A$ which satisfies $\|e\| = 1$.

From (1.1.1) there results $\|a^n\| \leqslant \|a\|^n$ so that we can easily verify that for any $a \in A$ with $\|a\| < 1$, $(e - a)^{-1}$ exists.

A complex homomorphism of A is a linear and multiplicative map from A into C.

Proposition 1.1. *The set $\mathfrak{M}$ of complex nonzero homomorphisms of A is a weak*-compact subset of the unit sphere of A^*, A^* being the dual of A.*

Proof. Let $\varphi \in \mathfrak{M}$. $\varphi \neq 0$ implies $\varphi(e) \neq 0$ and $\varphi(e) = \varphi(e)\,\varphi(e)$ implies $\varphi(e) = 1$. Let $a \in A$, $\|a\| \leqslant 1$ and $|\varphi(a)| > 1$. Setting $b = \dfrac{a}{\varphi(a)}$ we have $\|b\| < 1$, so that $(e - b)^{-1}$ exists.

Then

$$1 = \varphi(e) = \varphi[(e - b)(e - b)^{-1}] = \varphi(e - b)\,\varphi[(e - b)^{-1}] =$$
$$= [\varphi(e) - \varphi(b)]\,[\varphi(e - b)^{-1}] = 0,$$

which is not possible. It follows that $\|\varphi\| = 1$ for any $\varphi \in \mathfrak{M}$.

Since the unit sphere of A is weakly compact, it remains to show that $\mathfrak{M}$ is weakly closed.

Let $a^* \in A^*$ be a point in the weak*-closure of $\mathfrak{M}$, $\|a^*\| \leqslant 1$. For an $\varepsilon > 0$ and $a, b \in A$, let $V_{\varepsilon; a, b, ab}(a^*)$ be the neighbourhood of a^* defined by $V_{\varepsilon; a, b, ab}$.

We have $|a^*(a) - b^*(a)| < \varepsilon$, $|a^*(b) - b^*(b)| < \varepsilon$, $|a^*(ab) - b^*(ab)| < \varepsilon$ for any $b^* \in V_{\varepsilon; a, b, ab}(a^*)$. But a^* is in the weak*-closure of $\mathfrak{M}$ so that there exists a $\varphi \in \mathfrak{M}$ such that $\varphi \in V_{\varepsilon; a, b, ab}(a^*)$. Then

$$|a^*(ab) - a^*(a)\,a^*(b)| \leqslant |a^*(ab) - \varphi(ab)| + |a^*(a)|\,|\varphi(b) - a^*(b)| +$$
$$+ |\varphi(b)|\,|\varphi(a) - a^*(a)| < \varepsilon(1 + \|a\| + \|b\|).$$

Since ε is arbitrary, $a^*(ab) = a^*(a)\,a^*(b)$ for every $a, b \in A$, i.e. a^* is multiplicative on A. It remains to prove that $a^* \neq 0$. Suppose $a^* = 0$ and let $\varphi \in \mathfrak{M}$ such that $\varphi \in V_{1/2; e}(a^*)$. We have

$$1 = |a^*(e) - \varphi(e)| < 1/2$$

which is impossible. It follows that $a^* \neq 0$ so that $a^* \in \mathfrak{M}$, which completes the proof.

The compact Hausdorff space $\mathfrak{M}$ constructed this way is called the *maximal ideal space* of A. The elements of $\mathfrak{M}$ may be put into a one-to-one correspondence with the maximal ideals of A by the map $\varphi \to \operatorname{Ker} \varphi$, $\varphi \in \mathfrak{M}$.

Let $C(\mathfrak{M})$ be the Banach space of all continuous complex valued functions on $\mathfrak{M}$, with the supnorm. With the usual multiplication for functions, $C(\mathfrak{M})$ becomes a commutative Banach algebra with unit element.

For $a \in A$ we define on $\mathfrak{M}$ the function $\hat{a}$ by

$$\hat{a}(\varphi) = \varphi(a) \qquad (\varphi \in \mathfrak{M}). \tag{1.1.2}$$

It is easy to verify that $\hat{a} \in C(\mathfrak{M})$ and

$$\|\hat{a}\| = \sup_{\varphi \in \mathfrak{M}} |\hat{a}(\varphi)| \leqslant \|a\| \qquad (a \in A). \tag{1.1.3}$$

The map $a \to \hat{a}$ is a continuous algebra homomorphism from A into $C(\mathfrak{M})$. Let $\hat{A}$ be the range of this homomorphism.

Proposition 1.2. (*i*) *For any* $a \in A$ *we have*

$$\|\hat{a}\| = \lim \|a^n\|^{1/n},$$

(*ii*) *the homomorphism* $a \to \hat{a}$ *is injective if and only if*

$$\lim \|a^n\|^{1/n} = 0 \ \textit{ implies } \ a = 0 \qquad (a \in A),$$

(*iii*) *the homomorphism* $a \to \hat{a}$ *is a topological injection if and only if there exists a constant* K *such that*

$$\|a\|^2 \leqslant K\|a^2\| \qquad (a \in A),$$

(*iv*) *the homomorphism* $a \to \hat{a}$ *is an isometric injection if and only if*

$$\|a^2\| = \|a\|^2 \qquad (a \in A).$$

Proof. As is easily verified, $\hat{a}(\mathfrak{M})$ is just the spectrum of a so that (i) results from the spectral radius formula. The statement (ii) follows immediately from (i). We now prove (iii).

If $\|a\|^2 \leqslant K\|a^2\|$, then

$$\|a^{2^n}\|^{1/2^n} K^{1/2^n + \dots + 1/2^n} \geqslant \|a\|,$$

so that

$$K\|\hat{a}\| = K \lim_{n\to\infty} \|a^{2^n}\|^{1/2^n} \geqslant \|a\|.$$

Conversely, if $\|a\| \leqslant M\|\hat{a}\|$, then

$$\|a\|^2 \leqslant M^2\|\hat{a}^2\| \leqslant M^2\|a^2\|.$$

To prove (iv) we follow the above proof for (iii) exactly; we have for (iv) in addition: $K = M = 1$. The proof thus is complete.

In the case when all elements of A satisfy

$$\|a^2\| = \|a\|^2 \tag{1.1.4}$$

the homomorphism $a \to \hat{a}$ is an isometric embedding of A in $C(\mathfrak{M})$.

1.2. Measures

Let X be a compact Hausdorff space and $C(X)$ the Banach space of all continuous complex valued functions on X, with the supnorm. By a *measure on* X we mean a linear continuous functional on $C(X)$. We call a measure μ on X *real* if for any real valued $f \in C(X)$, $\mu(f)$ is real. The measure μ on X is called positive if for any positive function $f \in C(X)$, $\mu(f)$ is positive.

Let μ be a positive measure on X. The theory of integration relative to μ together with that of the spaces $L^p(\mathrm{d}\mu)$ is well known. For $1 \leqslant p < \infty$, $L^p(\mathrm{d}\mu)$ is a Banach space with the norm

$$[\int |f|^p \,\mathrm{d}\mu]^{1/p}.$$

$L^\infty(\mathrm{d}\mu)$ is a Banach algebra, with the ess supnorm.

If $f \in L^p(\mathrm{d}\mu)$ and $g \in L^q(\mathrm{d}\mu)$ with $1/p + 1/q = 1$, then $fg \in L^1(\mathrm{d}\mu)$ and Hölder inequality holds:

$$|\int fg d\mu| \leqslant [\int |f|^p \, d\mu]^{1/p} [\int |g|^q \, d\mu]^{1/q}.$$

The equality sign holds if and only if f and g are linearly dependent.

For $p \geqslant 1$, any function in $L^p(\mathrm{d}\mu)$ is also in $L^1(\mathrm{d}\mu)$. The set of all functions of $C(X)$ forms (modulus $\mathrm{d}\mu$) a dense subspace in $L^p(\mathrm{d}\mu)$ for $1 \leqslant p < \infty$, and a weakly*-dense subspace in $L^\infty(\mathrm{d}\mu)$.

For $1 \leqslant p < \infty$ and $1 < q \leqslant \infty$ with $1/p + 1/q = 1$, the dual space of $L^p(\mathrm{d}\mu)$ is isometrically isomorphic to $L^q(\mathrm{d}\mu)$.

The characteristic function χ_ω of a Borel set $\omega \subset X$ belongs to $L^p(\mathrm{d}\mu)$. We put

$$\mu(\omega) = \int \chi_\omega \mathrm{d}\mu.$$

We say that the positive measure μ has a support in the Borel set ω if $\mu(X - \omega) = 0$.

Any measure μ can be uniquely written as $\mu = \mu_1 + i\mu_2$, μ_1, μ_2 being real measures. For any real measure μ one can uniquely define the positive measures μ^+ and μ^- with supports in two disjoint Borel sets and such that $\mu = \mu^+ - \mu^-$.

For a measure μ of the form $\mu = \mu_1 + i\mu_2$ and a Borel set ω, we put

$$\mu(\omega) = \mu_1^+(\omega) - \mu_1^-(\omega) + i[\mu_2^+(\omega) - \mu_2^-(\omega)].$$

For any measure μ there exists a unique positive measure $|\mu|$ and a uniquely determined function $F \in L^1(d\mu)$, such that $|F| = 1$ $|\mu|$-almost everywhere and $d\mu = Fd|\mu|$. We have $\|\mu\| = |\mu|(1)$. If μ is real, then $|\mu| = \mu^+ + \mu^-$. A measure μ is positive if and only if $\|\mu\| = \mu(1)$.

We say that a measure μ_1 is *absolutely continuous* with respect to the measure μ, if for any Borel set $\omega \subset X$ with $\mu(\omega) = 0$ we have $\mu_1(\omega) = 0$.

If μ_1 is absolutely continuous with respect to the positive measure μ, then there exists a function $f \in L^1(d\mu)$ such that $d\mu_1 = fd\mu$ (Radon-Nikodym theorem).

The measures μ_1 and μ_2 are called mutually-*singular* if there exist two disjoint Borel sets ω_1 and ω_2 such that for any Borel set ω one has $\mu_1(\omega \cap \omega_1) = \mu_1(\omega)$, $\mu_2(\omega \cap \omega_2) = \mu_2(\omega)$.

If μ is a positive measure and μ_1 an arbitrary one, then μ_1 can be uniquely written in the form $\mu_1 = \mu_a + \mu_s$, where μ_a is absolutely continuous with respect to μ and μ_s and μ are mutually-singular (Lebesgue's decomposition theorem).

Let K be a closed subset of X.

Any measure μ on K can be considered as a measure on X by putting $\mu(f) = \mu(f/K)$ for $g \in C(X)$. If μ is a measure on X we put $\mu_K = \chi_K\mu$.

We note that the space $\mathcal{M}(X)$ of all measures on X is the dual of the Banach space $C(X)$.

1.3. Convexity

Let B be a real normed space and B^* its dual space. The space B^* endowed with the weak*-topology is a locally convex space and B is (isomorphic to) its dual.

A subset E of B^* is called *convex* if it is non-empty and for each $a, b \in E$ the segment $[a, b] = \{c \in B^*: c = ta + (1 - t)b, 0 \leqslant t \leqslant 1\}$ is included in E. It is clear that any non-empty intersection of convex sets is a convex set. Thus for a non-empty set $F \subset B^*$ there is a smallest closed convex set which contains F. We denote it by $\langle F \rangle$ and call it the *closed convex hull* of F.

The set of all convex combinations of elements from F is evidently convex, so its closure contains $\langle F \rangle$. The converse is clearly also true, thus $\langle F \rangle$ is the closure of all convex combinations of the elements of F.

Proposition 1.3. *Let K be a compact subset of B^*. The closed convex hull of K is compact and we have*

$$\langle K \rangle = \{b \in B^*: b(u) \leqslant \sup_{k \in K} k(u); u \in B\}.$$

Let E be a convex set in B^*. We call $e \in E$ an *extremal point* of E if there is no segment of E containing e in its interior.

Proposition 1.4. (Krein-Milman theorem). *Let K be a convex compact set in B^*. Then K is the closed convex hull of the set of its extremal points.*

1.4. Holomorphic functions of several complex variables

We shall denote by C^n the cartesian product $C \times C \times \ldots \times C$ of n copies of the complex plane C.

The points in C^n will be written as $z = (z_1, \ldots, z_n)$. If $z \in C^n$ then we put

$$|z| = \max \{|z_j|; \ 1 \leqslant j \leqslant n\}.$$

Let $w \in C^n$ and $r = (r_1, \ldots, r_n) \in R^n$, $r_j > 0$.

By an open *polydisk* of *polyradius* r and center w we mean the set $\Delta(w, r)$ defined by

$$\Delta(w; r) = \{z \in C^n: |z_j - w_j| < r_j, 1 \leqslant j \leqslant n\}.$$

Let D be an open set in C^n. A function f defined on D and with values in C will be called holomorphic in D if for any $w \in D$ there exists a neighbourhood U of w such that $w \in U \subset D$ and for any $z \in U$

$$f(z) = \sum_{k_1, k_2, \dots, k_n = 0}^{\infty} a_{k_1, k_2, \dots, k_n}(z_1 - w_1)^{k_1} \dots (z_n - w_n)^{k_n}$$

the series being convergent for every $z \in U$.

Let $P_1, P_2, \dots, P_r$ be a system of polynomials in n variables $z_1, z_2, \dots, z_n$. The *polynomial polyhedron* of radius δ defined by these polynomials is the open set

$$Q(P_1, \dots P_r; \delta) = \{z \in C^n \colon |z| < \delta,\ 1 \leqslant i \leqslant n,$$

$$|P_j(z)| < \delta, 1 \leqslant j \leqslant r\}.$$

Let D be an open set in C^n. A *system of Cousin data* on D is the system $\{U_i, h_{ij}\}$, where $\{U_i\}$ forms a finite open covering of D, that is U_i are open sets $U_i \subset D$ with $D = \bigcup U_i$ and h_{ij} are functions defined and holomorphic on $U_i \cap U_j$, which satisfy

$$h_{ij} + h_{ji} = 0 \qquad \text{on } U_i \cap U_j$$

$$h_{ij} + h_{jk} + h_{ki} = 0 \qquad \text{on } U_i \cap U_j \cap U_k.$$

Proposition 1.5. (Cousin's theorem). *Let $\{U_i, h_{ij}\}$ be a system of Cousin data on a polynomial polyhedron Q. Then there is a system of functions h_i so that h_i is defined and holomorphic on U_i and*

$$h_i - h_j = h_{ij} \qquad \text{on } U_i \cap U_j.$$

For a compact set K of C^n we define the *polynomial-convex hull* of K as

$$\hat{K} = \{z \in C^n \colon |P(z)| \leqslant \max_{\omega \in K} |P(\omega)| \text{ for any polynomial } P\}.$$

The compact set K is called *polynomially convex* if $\hat{K} = K$.

Proposition 1.6. (Oka's theorem). *Let K be a compact polynomially convex set of* C^n. *Any holomorphic function on a neigbourhood of K is a uniform limit on K of polynomials.*

Notes

The purpose of this chapter is to define the basic concepts and notations of the elements of commutative Banach algebras, the measure theory, the convexity theory and the theory of holomorphic functions of several complex variables, used throughout this book.

The present chapter also contains a number of mathematical results which help readers to quickly gain the required information for the following chapters.

Detailed expositions on the mathematical facts presented in this introductory chapter can be found in: I. Gelfand, D. Raikov, G. Shilov, [1] and C. Foiaş [4] for commutative Banach algebras, N. Bourbaki [1], N. Dinculeanu [1] and M. Nicolescu [1] for measure theory, N. Bourbaki [2], N. Dunford J. Schwartz [1] and K. Yosida [1] for convexity theory and other elements of functional analysis, and E. Gunning, H. Rossi [1] for holomorphic functions of several complex variables.

CHAPTER 2

Boundaries

2.1. Function algebras

Let X be a compact Hausdorff space.

We shall denote by $C(X)$ the Banach algebra of all complex valued continuous functions on X, provided with the supremum norm

$$\|f\| = \sup_{x \in X} |f(x)| \qquad (f \in C(X)).$$

By $C_R(X)$ we denote the real Banach algebra of the real valued functions belonging to $C(X)$.

A subalgebra A of $C(X)$ will be called a *function algebra* on X if the following conditions are satisfied:

a) A is uniformly closed in $C(X)$,

b) A separates the points of X: for any $x_1, x_2 \in X$, $x_1 \neq x_2$, there is an $f \in A$ such that $f(x_1) \neq f(x_2)$,

c) A contains the constant functions.

Any function algebra is a Banach algebra with unit element $e = 1$.

For a function $f \in C(X)$ we denote by $\mathrm{Re}f$ its real part and by $\bar{f}$ its complex conjugate. We also use the following notations

$$\mathrm{Re}A = \{u \in C_R(X) : u = \mathrm{Re}f \quad \text{with } f \in A\},$$

$$\bar{A} = \{g \in C(X) : g = \bar{f} \qquad \text{with } f \in A\}.$$

The set ReA is a vector subspace of $C_R(X)$ and $\bar{A}$ is a function algebra on X.

A very simple example of a function algebra on X is $C(X)$. Another is furnished by the most appropriate model of the entire theory: let $X = \{z \in C: |z| = 1\}$ be the unit circle of the complex plane and A the algebra of continuous functions on X which have an analytic extension into the interior of the unit disc. A is a function algebra on X and we have $A \neq C(X)$. In the following we shall call this algebra the *standard algebra*.

2.2. Representing measures

Let A be a function algebra on X. A positive measure μ on X will be called a *representing measure* for the point $x \in X$ if

$$f(x) = \int f \mathrm{d}\mu \qquad (f \in A). \tag{2.2.1}$$

It follows from (2.2.1) that representing measures are multiplicative on A and that $\mu(X) = 1$.

A positive measure μ on X is a representing measure for $x \in X$ if and only if

$$u(x) = \int u \mathrm{d}\mu \qquad (u \in \mathrm{Re}A). \tag{2.2.2}$$

The point mass ε_x defined by

$$\varepsilon_x(f) = f(x) \qquad (f \in C(X))$$

is clearly a representing measure for x. To simplify the notation we shall write $\mu \ll \varepsilon_x$ for "μ is a representing measure for x."

Let us fix an $x \in X$ and let S be a closed subset of X such that for any $u \in \mathrm{Re}A$, $u \geqslant 0$ on S, we have $u(x) \geqslant 0$.

For a function $v \in C_R(S)$ we define

$$Q_x^S(v) = \inf \{u(x): u \in \mathrm{Re}A,\ u \geqslant v \text{ on } S\}.$$

One easily verifies the following properties of the functional Q_x^S on $C_R(S)$:

a) $Q_x^S(v_1 + v_2) \leqslant Q_x^S(v_1) + Q_x^S(v_2)$,

b) $Q_x^S(\alpha v) = \alpha Q_x^S(v)$ for $\alpha \geqslant 0$,

c) $v_1 \leqslant v_2$ implies $Q_x^S(v_1) \leqslant Q_x^S(v_2)$,

d) $Q_x^S(u) = u(x)$ for any $u \in \mathrm{Re}A$,

e) $-Q_x^S(-v) = \sup\{u(x),\ u \in \mathrm{Re}A,\ u \leqslant v \text{ on } S\}$.

Theorem 2.1. *Let $x \in X$ and $S \subset X$, closed, such that for any function u in ReA, non-negative on S, we have $u(x) \geqslant 0$. For any $w \in C_R(S)$ and any α belonging to the closed interval $[-Q_x^S(-w), Q_x^S(w)]$ there exists a measure $\mu \ll \varepsilon_x$ so that $\mu(X - S) = 0$ and $\mu(w) = \alpha$. Conversely, for any $\mu \ll \varepsilon_x$ with $\mu(X - S) = 0$, we have $\mu(w) \in [-Q_x^S(-w), Q_x^S(w)]$.*

Proof. As the functional Q_x^S is subadditive, positively homogeneous and bounded on $C_R(S)$, by applying the Hahn-Banach theorem we find a real measure μ on S such that $\mu(v) \leqslant Q_x^S(v)$ for any $v \in C_R(S)$ and $\mu(w) = \alpha$. But $Q_x^S(v) \leqslant 0$ for $v \leqslant 0$, hence $\mu(v) \leqslant 0$ for $v \leqslant 0$ and μ is a positive measure on S.

For $u \in \mathrm{Re}A$ we have

$$\mu(u) \leqslant Q_x^S(u) = u(x)$$

$$\mu(-u) \leqslant Q_x^S(-u) = -u(x)$$

which means that $\mu(u) = u(x)$ and thus $\mu \ll \varepsilon_x$. The measure μ may be considered as a measure on X and we have $\mu(X - S) = 0$.

The second part of the theorem follows immediately from the definition of Q_x^S and property (e).

The proof is complete. ◇

Corollary 2.2. *Let x and S be as in Theorem 2.1. The necessary and sufficient condition for x to admit a unique representing measure with support contained in S is that for any $v \in C_R(S)$, $-Q_x^S(-v) = Q_x^S(v)$ i.e.*

$$\sup\{u(x) : u \in \mathrm{Re}A,\ u \leqslant v \text{ on } S\} = \inf\{u(x) : u \in \mathrm{Re}A,\ u \geqslant v \text{ on } S\}.$$

Corollary 2.3. *Let x and S be as in Theorem 2.1. Then for any* $v \in C_R(S)$

$$Q_x^S(v) = \sup \{\mu(v) : \mu \ll \varepsilon_x, \mu(X - S) = 0\}.$$

2.3. The Choquet boundary

A non-empty closed set $E \subset X$ will be called *absorbant* if for any $x \in E$ and $\mu \ll \varepsilon_x$ we have $\mu(X - E) = 0$.

Proposition 2.4. *Any closed set which is a union of absorbant sets is an absorbant set. Any non-empty intersection of absorbant sets is absorbant.*

Proof. The first statement is obvious. To prove the second one it is sufficient to note that if F is a closed set, such that for any open set U with $F \subset U$ there is an absorbant set E with the property $F \subset E \subset U$, then F is absorbant.

Hence the family of all absorbant sets is inductively ordered by inclusion and from Zorn's lemma we deduce that any absorbant set contains a minimal absorbant one. Since X is clearly an absorbant set, *the existence of minimal absorbant sets results.*

Proposition 2.5. *Let E be an absorbant set and* $u \in ReA$, *such that* $u \geqslant 0$ *on E. Then, the set* $E_1 = \{x \in E;\ u(x) = 0\}$ *is either absorbant or empty.*

Proof. Let $x \in E_1$ and $\mu \ll \varepsilon_x$. Let K be an arbitrary compact set in $X - E_1$. Since $u > 0$ on $K \cap E$, there exists a positive constant r such that $ru \geqslant 1$ on $K \cap E$. We have

$$\mu(K) = \mu(K \cap E) + \mu((X - E) \cap K) = \mu(K \cap E) \leqslant$$

$$\leqslant \mu(ru) = r\mu(u) = ru(x) = 0.$$

A non-empty closed set $K \subset X$ is called a *peak set* (with respect to A) if there is a function $f \in A$ with the properties: $f(x) = 1$ for any $x \in K$ and $|f(x)| < 1$ for any $x \in X - K$. The point x is said to be a *peak point* if $\{x\}$ is a peak set.

Proposition 2.6. *For any peak set K there exists a function $u \in ReA$, $u \geqslant 0$, such that $K = \{x \in X: u(x) = 0\}$. If $u \in ReA$, $u \geqslant 0$ and $F = \{x \in X: u(x) = 0\}$, then for any $y \in F$ there is a peak set K such that $y \in K \subset F$.*

Proof. Let K be a peak set and $f \in A$ be such that $f(x) = 1$ for $x \in K$ and $|f(x)| < 1$ for $x \in X - K$. If $g = 1 - f$ and $u = \text{Re}g$ then $u \in \text{Re}A$, $u \geqslant 0$ and $K = \{x \in X: u(x) = 0\}$.

Now let $u \in \text{Re}A$, $u \geqslant 0$ and $F = \{x \in X: u(x) = 0\}$. Take $g \in A$ such that $u = \text{Re}g$ and put $h = e^{-g}$. Then we get $|h(x)| = 1$ for $x \in F$ and $|h(x)| < 1$ for $x \in X - F$. For a fixed $y \in F$ let $f = \dfrac{1}{2}(e^{-i\theta} h + 1)$, where $\theta = \arg h(y)$. We have $f \in A$, $|f| \leqslant 1$, $f(y) = 1$, $f(x) = 1$ if and only if $|f(x)| = 1$. If we put $K = \{x \in X: f(x) = 1\}$ then K is a peak set and $y \in K \subset F$.

Proposition 2.7. *A non-empty intersection of peak sets is absorbant. If a non-empty G_δ-set is an intersection of peak sets, then it is a peak set.*

Proof. From Proposition 2.5 and 2.6, it follows that any peak set is absorbant. Hence the first statement follows from Proposition 2.4.

Let $F = \bigcap U_n$ be a non-empty G_δ-set which is an intersection of peak sets. Then it clearly results that for any n there is a peak set K_n such that $F \subset K_n \subset U_n$. We get then $F = \bigcap K_n$. Let $f_n \in A$ with $f_n(x) = 1$ for any $x \in K_n$ und $|f_n(x)| < 1$ for $x \in X - K_n$, and

$$f = \sum_{n=1}^{\infty} \frac{1}{2^n} f_n.$$

We have $f \in A$, $f(x) = 1$ for $x \in F$ and $|f(x)| < 1$ for $x \in X - F$, that is F is a peak set.

Theorem 2.8. *Let $x \in X$. The following statements are equivalent:*

a) *x belongs to a minimal absorbant set;*

b) *$\{x\}$ is an absorbant set;*

c) *ε_x is the only representing measure for x;*

d) *For any neighbourhood U of x and any positive real numbers r_1, r_2 there is a $u \in \text{Re}A$, $u \geqslant 0$, such that $u(x) < r_1$ and $u \geqslant r_2$ on $X - U$;*

e) *For any closed set F of G_δ-type, containing x, there is a peak set K such that $x \in K \subset F$;*

f) *For any neighbourhood U of x there exists a peak set K with the property $x \in K \subset U$;*

g) *For any* $y \neq x$ *there is a* $u \in \mathrm{Re}A$ *so that* $u \geqslant 0$ *on* X, $u(x) = 0$ *and* $u(y) > 0$;

h) $\{x\}$ *is an intersection of peak sets.*

Proof. (a) → (b). Let E be a minimal absorbant set with $x \in E$. We show that for any $f \in A$ and $y \in E$ we have $f(y) = f(x)$ and, since A separates the points of X it will result that $\{x\} = E =$ absorbant. It is sufficient to prove that for any $u \in \mathrm{Re}A$ and $y \in E$ we have $u(y) = u(x)$. Obviously, one can also suppose $u > 0$.

Let

$$r = \inf\left\{\frac{u(x)}{u(y)}; \quad y \in E\right\}.$$

E is compact so that there is an $x_0 \in E$ with

$$r = \frac{u(x)}{u(x_0)}.$$

Let $u_0 = u(x) - ru$. We have $u_0 \in \mathrm{Re}A$, $u_0 \geqslant 0$ on E. Let us write

$$E_0 = \{y \in E\colon u_0(y) = 0\}.$$

Clearly $x_0 \in E_0$.

By Proposition 2.5 one gets that E_0 is absorbant. But E is a minimal absorbant set, hence $E_0 = E$. Then, for any $y \in E$ we have $u(x) = ru(y)$ and, since in particular the same holds for x, $r = 1$ and $u(y) = u(x)$ for any $y \in E$.

(b) → (c) is immediate.

(c) → (d). Let U be a neighbourhood of x, $v \in C_R(X)$, $v \geqslant 0$, such that $v(x) = 0$ and $v \geqslant r_2$ on $X - U$. By use of Corollary 2.2 one gets from (c)

$$0 = v(x) = \inf\{u(x)\colon u \in \mathrm{Re}A,\ u \geqslant v\}$$

so that there exists a $u \in \mathrm{Re}A$ as required in (d).

(d) → (e). Let F be a closed G_δ-set, $F = \bigcap V_n$, with $x \in F$. Using induction we construct a sequence of open sets U_n and a sequence of functions g_n from A such that

$$\text{(i)}\ \ \|g_{n+1} - g_n\| \leqslant 2^{-n+1}$$

$$\text{(ii)}\ \ \|g_n\| \leqslant 3(1 - 2^{-n-1})$$

(iii) $g_n(x) = 3(1 - 2^{-n})$

(iv) $|g_{n+1} - g_n| < 2^{-n-1}$ on $X - U_{n+1}$

(v) $x \in U_{n+1} \subset V_n$

Taking $r_1 = -\ln(1 - \varepsilon)$, $r_2 = -\ln\varepsilon$, it results from (d) that for any open U with $x \in U$, there is an $f \in A$ such that $\|f\| \leqslant 1$, $|f(x)| > 1 - \varepsilon$ and $|f| < \varepsilon$ on $X - U$.

Let $f \in A$ be such that

$$\|f\| \leqslant 1,\ |f(x)| > \frac{3}{4} \text{ and } |f| < \frac{1}{4} \text{ on } X - V_1.$$

We put

$$g_1 = \frac{3}{2}[f(x)]^{-1} f \text{ and } U_1 = V_1.$$

We have

$$\|g_1\| \leqslant \frac{3}{2} \cdot \frac{4}{3} = 2 < 3(1 - 2^{-2})$$

hence g_1 satisfies (ii). At the same time

$$g_1(x) = \frac{3}{2} = 3(1 - 2^{-1})$$

so g_1 satisfies (iii). Suppose we have already constructed $U_1, \ldots, U_k$ and $g_1, \ldots, g_k$ with the properties (i) — (v). Let

$$U = \{y \in X: |g_k(y)| < 3(1 - 2^{-k}) + 2^{-k-2}\}.$$

Since $g_k(x) = 3(1 - 2^{-k})$, we have $x \in U \cap V_k$. We now write $U_{k+1} = U \cap V_k$. Then $x \in U_{k+1} \subset V_k$, i.e.(v). Let $f \in A$ with $|f| < 1$,

$$|f(x)| > \frac{3}{4} \text{ and } |f| < \frac{1}{4} \text{ on } X - U_{k+1},$$

and write $h = 3.2^{-k-1}[f(x)]^{-1} f$.

We have $h(x) = 3.2^{-k-1}$, $\|h\| \leqslant 2^{-k+1}$ and, for $y \in X - U_{k+1}$

$$|h(y)| = 3.2.^{k-1}[f(x)]^{-1}|f(y)| \leqslant 3.2^{-k-1}\frac{4}{3}\cdot\frac{1}{4} = 2^{-k-1}.$$

Let $g_{k+1} = g_k + h$. From (i) — (v) for g_k and from the properties of h we get

$$\|g_{k+1} - g_k\| = \|h\| \leqslant 2^{-k+1},$$

$$g_{k+1}(x) = g_k(x) + h(x) = 3(1 - 2^{-k}) + 3.2^{-k+1} = 3(1 - 2^{-(k+1)})$$

$$|g_{k+1} - g_k| = |h| < 2^{-k-1} \text{ on } X - U_{k+1}$$

which are just (i), (iii) and (iv).

For $y \in X - U_{k+1}$ we have

$$|g_{k+1}(y)| \leqslant \|g_k\| + |h(y)| \leqslant 3(1 - 2^{-k-1}) + 2^{-k-1} =$$

$$= 3(1 - 2.2^{-k-2}) + 2.2^{-k-2} \leqslant 3(1 - 2^{-k-2}).$$

For $y \in U_{k+1}$ we have $y \in U$, hence

$$|g_{k+1}(y)| \leqslant |g_k(y)| + \|h\| < 3(1 - 2^{-k}) + 2^{-k-2} + 2^{-k+1} =$$

$$= 3(1 - 4.2^{-k-2}) + 2^{-k-2} + 8.2^{-k-2} = 3(1 - 2^{-k-2}).$$

Then, it results that

$$\|g_{k+1}\| \leqslant 3(1 - 2^{-k-2}),$$

i.e. property (ii) for g_{k+1}.

Thus, the sequence g_n with properties (i)—(v) has been constructed. (i) implies that the series

$$\sum_{n=1}^{\infty} (g_{n+1} - g_n)$$

is uniformly convergent, hence

$$g = \sum_{n=1}^{\infty} (g_{n+1} - g_n) \in A.$$

From (ii) and (iii) it results that $\|g\| = 3$ and from (iv) and (v) we get

$$|g(y)| \leqslant \|g_n\| + \sum_{k=n}^{\infty} |g_{k+1}(y) - g_k(y)| < 3\,(1 - 2^{-n-1}) + \sum_{k=n}^{\infty} 2^{-k-1} =$$

$$= 3 - 1/(2^{n+1}) < 3,$$

for any $y \in X - F$. Writing $f = \dfrac{1}{6}(3 + g)$ and $K = f^{-1}\,(1)$, we see that K is a peak set and $x \in K \subset F$.

(e) → (f). Let U be a neighbourhood of x and $v \in C_R(X)$, $v \geqslant 0$, $v(x) = 0$ and $v \geqslant 1$ on $X - U$. Then $F = \{x : v(x) = 0\}$ is a closed set of G_δ-type and $x \in F \subset U$.

(f) → (g) follows immediately.

From (g) we have

$$\{x\} = \bigcap_{u \in \delta_x} \{y \in X : u(y) = 0\},$$

where $\delta_x = \{u \in \operatorname{Re}A : u \geqslant 0,\ u(x) = 0\}$. Then Proposition 2.6 implies that $\{x\}$ is an intersection of peak sets.

(h) → (b) is proved by Proposition 2.7 and (b) → (a) is obvious.

The proof of Theorem 2.8 is complete. ◇

A point $x \in X$ will be called a *Choquet point* if it satisfies one of the (equivalent) assertions (a) — (h) of Theorem 2.8. The set of all Choquet points of X is called the *Choquet boundary* of X (relative to A). We denote it by $\Sigma = \Sigma(A; X)$.

It will be proved in the following that the Choquet boundary depends only on the Banach algebra structure of A.

Point (a) of Theorem 2.8 shows that Σ is non-empty.

Corollary 2.9. *Any intersection of peak sets contains a Choquet point.*

Corollary 2.10. *For any $f \in A$, there is an $x \in \Sigma$ such that $|f(x)| = \|f\|$.*

Proof. Suppose $\|f\| = 1$ and let $g = \frac{e^{it}}{2}(1 + f)$ where t is chosen such that the set $K = \{x; g(x) = 1\}$ is non-empty. Then K is a peak set and hence it contains a Choquet point. But $|f| = 1$ on K, so that the Corollary is proved.

Corollary 2.11. *Any Choquet point with countable basis is a peak point.*

Corollary 2.12. *If any Choquet point has countable basis then Σ is the smallest subset of X on which, for any $f \in A$, $|f|$ attains its maximum.*

2.4. The Shilov boundary

A closed subset S of X will be called a *determining set* if for any $u \in \text{Re}A$, $u \geqslant 0$ on S, we have $u \geqslant 0$ on X.

Proposition 2.13. *A closed set S is a determining set if and only if for any $f \in A$ there is an $x \in S$ such that $|f(x)| = \|f\|$.*

Proof. Let S be a determining set, $f \in A$ and $m = \sup \{|f(x)|; x \in S\}$. Let $y \in X$, $\theta = \arg f(y)$ and $g = m - e^{-i\theta} f$. The function $u = \text{Re} g$ is obviously positive on S, hence $u \geqslant 0$ on X. Then, $u(y) = m - e^{-i\theta} f(y) \geqslant \geqslant 0$, so that

$$|f(y)| = e^{-i\theta} f(y) \leqslant m.$$

Conversely, assume that the modulus of any function in A attains its maximum (on X) on the closed set S. Let $u \in \text{Re}A$, $u \geqslant 0$ on S and $f = u + iv \in A$. If we put $g = e^{-f}$ then $g \in A$ and $|g| = e^{-u} \leqslant 1$ on S. There results $e^{-u} = |g| \leqslant 1$ on X, thus $u \geqslant 0$ on X.

Proposition 2.14. *For any absorbant set E and any determining set S we have $E \cap S \neq \varnothing$.*

Proof. If $E \cap S = \varnothing$, then for any $x \in E$ and $\mu \lll \varepsilon_x$ one has $\mu(S) = 0$. On the other hand, the condition of Theorem 2.1 is fulfilled for S and any point $x \in E$, hence there exists a $\mu \lll \varepsilon_x$ with support in S, which contradicts the previous assertion. ◇

Corollary 2.15. *Any determining set contains the Choquet boundary.*

From Corollary 2.10 it follows that the closure of the Choquet boundary is determining and from Corollary 2.15 it follows that there exists a smallest determining set.

The smallest determining set is called the *Shilov boundary* of X relative to A and is denoted by $\Gamma = \Gamma(X, A)$.

Theorem 2.16. *The Shilov boundary is equal to the closure of the Choquet boundary. A point $x \in X$ belongs to the Shilov boundary if and only if for any neighbourhood U of x there is a function $f \in A$ such that $|f| < \|f\|$ on $X - U$.*

Proof. Since the closure of the Choquet boundary is determining, it contains the Shilov boundary. Then, from Corollary 2.15 there results that the closure of the Choquet boundary is simply the Shilov boundary. The second part of the theorem follows from point (f) of Theorem 2.8.

In the case of the algebra $C(X)$, the Choquet boundary is equal to the Shilov boundary and equal to X; this follows immediately from point (c) of Theorem 2.8.

Let A be the standard algebra on $X = \{z : |z| = 1\}$ and $\mu \preccurlyeq \varepsilon_x$, $x \in X$. Then, the real measure $\nu = \mu - \varepsilon_x$ satisfies the relation

$$\int z^n d\nu = 0 \qquad (n = 0, +1, \ldots)$$

hence $\nu = 0$. It follows that $\mu = \varepsilon_x$ and point (c) of Theorem 2.8 yields

$$\Sigma = \Gamma = X.$$

The Choquet boundary is generally not closed, and is hence different from the Shilov boundary. To illustrate this point, we shall give an example.

Let $X = \{z \in C : |z| = 1\}$ and A the algebra of continuous functions f on X, having analytic extension into the interior of the unit disc and verifying $f(0) = f(1)$.

Since X is the Choquet boundary of the standard algebra on X, from Corollary 2.11 it follows that for any $x \in X$ there is a continuous function f_x on X, having analytic extension in the interior of the unit disc, such that $f_x(x) = 1$ and $|f_x| < 1$ on $X - \{x\}$. Let $x \in X$, $x \neq 1$ and let $g \in A$ be defined by

$$g = z \cdot \frac{f_x - f_x(1)}{1 - \overline{f_x(1)} f_x}.$$

Obviously, $g \in A$, $|g(x)| = 1$, $|g| < 1$ on $X - \{x\}$. Thus x is a peak point for A, hence a Choquet point for A, and this is true for any $x \in X$, $x \neq 1$. If μ is the normalized Lebesgue measure on X, then it is known that

$$f(0) = \int f d\mu \qquad (f \in A)$$

and as $f(0) = f(1)$ for $f \in A$, we have $\mu \ll \varepsilon_1$. But clearly $\mu \neq \varepsilon_1$, hence from point (c) of Theorem 2.8, 1 does not belong to the Choquet boundary of X relative to A. Then, in this case, we have $\Sigma = X - \{1\}$ and $\Gamma = X$; Σ is not closed, and is hence different from Γ.

Now let A be a function algebra on a compact Hausdorff space X and $F \subset X$, F closed. We denote by $A \mid F$ the algebra of all functions in $C(F)$ which are restrictions to F of functions in A.

Theorem 2.17. *Let $S \subset X$ be a determining set. Then: (i) the algebra $A|S$ is a function algebra on S, isometrically isomorphic to A;*

(ii) a closed set $K_1 \subset S$ is a peak set for $A|S$ if and only if there exists a peak set $K \subset X$ for A such that $K_1 = K \cap S$;

(iii) The Choquet (Shilov) boundary of S relative to $A \mid S$ is equal to the Choquet (Shilov) boundary of X relative to A.

Proof. (i) Let T be the restriction operator defined on A, with values in $A \mid S$

$$Tf = f|S \qquad (f \in A).$$

T is linear and maps the algebra A onto the algebra $A \mid S$. Since S is a determining set, we have $\|Tf\| = \|f\|$ hence T is isometric. Thus, T is an isometric isomorphism between A and $A|S$, which gives that $A \mid S$ is a function algebra on S.

(ii) Let $K \subset S$ be a peak set for $A \mid S$ and $u \in \mathrm{Re}A$ be such that $u \geqslant 0$ on S and $K_1 = \{x \in S: u(x) = 0\}$.

Take $K = \{x \in X; u(x) = 0\}$. As S is a determining set, we have $u \geqslant 0$ on X and K results a peak set for A. Clearly $K_1 = K \cap S$.

The converse statement is obvious.

(iii) Since S is determining, we get $\Sigma(A; X) \subset S$.

Suppose $x \in S$ is an intersection of peak sets for A, $\{x\} = \bigcap K_i$. Then one has $\{x\} = \{x\} \cap S = \bigcap(K_i \cap S)$ and from (ii) there results that x is an intersection of peak sets for $A|S$. Thus, we get $\Sigma(A; X) \subset \subset \Sigma(A \mid S; S)$.

Conversely, let $x \in \Sigma(A \mid S, S)$ and let K be the intersection of all peak sets for A, containing x. (ii) gives $S \cap K = \{x\}$. Since K is an intersection of peak sets for A, it results from Proposition 2.7 that K is absorbant for A. Let E be a minimal absorbant set for A, $E \subset K$. From Proposition 2.14 we have $E \cap S \neq \varnothing$. Then

$$\varnothing \neq E \cap S \subset K \cap S = \{x\}.$$

Hence $x \in E$ and point (a) of Theorem 2.8 gives $x \in \Sigma(A, X)$.

2.5. Geometric characterization

Let A be a function algebra on X and A^* the dual of the Banach space A. When endowed with the weak*-topology, A^* becomes a locally convex space.

For an $x \in X$, let e_x be the functional defined on A by

$$e_x(f) = f(x).$$

It is clear that $e_x \in A^*$ and that the map $x \to e_x$ is a homeomorphic embedding of X into A^*. In this sense, we shall consider X as a compact subset of A^*.

In the following, we shall consider A^* as a locally convex space over the field of real numbers. Then its dual is ReA.

We recall that for $F \subset A^*$ we denote by $\langle F \rangle$ its closed convex hull.

Proposition 2.18. *The closed convex hull* $\langle X \rangle$ *of* X *is given by*

$$\langle X \rangle = \{a^* \in A^*: a^*(1) = \|a^*\| = 1\}.$$

Proof. From Proposition 1.3 we get

$$\langle X \rangle = \{a^* \in A^*: a^*(u) \leqslant \sup_{x \in X} u(x),\ u \in \mathrm{Re}A\}.$$

Let $a^* \in \langle X \rangle$. We have

$$a^*(u) \leqslant \sup_{x \in X} u(x) \leqslant \|u\| \qquad (u \in \mathrm{Re}A),$$

$$-a^*(u) = a^*(-u) \leqslant \sup_{x \in X} (-u(x)) \leqslant \|u\| \qquad (u \in \mathrm{Re}A).$$

Therefore

$$-\|u\| \leqslant a^*(u) < \|u\| \qquad (u \in \mathrm{Re}A),$$

i.e. $\|a^*\| \leqslant 1$.

On the other hand, $-a^*(1) = a^*(-1) \leqslant \sup(-1) = -1$. Hence $a^*(1) \geqslant 1$, that is

$$a^*(1) = 1 = \|a^*\|.$$

Conversely, let $a^* \in A^*$ be such that

$$a^*(1) = 1 = \|a^*\|.$$

Suppose $a^* \notin \langle X \rangle$, which means that there is a $u \in \mathrm{Re}A$ such that $a^*(u) > \sup_{x \in X} u(x)$. Let n be large enough to ensure that $n + u \geqslant 0$ on X. Then $\|n + u\| = n + \sup_{x \in X} u(x)$. We have

$$a^*(n + u) = n + a^*(u) > n + \sup_{x \in X} u(x) \geqslant \|n + u\|$$

which contradicts the fact that $\|a^*\| = 1$.

Proposition 2.19. *Let $a^* \in A^*$. Then $a^* \in \langle X \rangle$ if and only if there exists a positive measure μ on X, $\mu(X) = 1$, such that*

$$a^*(f) = \int f \mathrm{d}\mu \qquad (f \in A).$$

Proof. If there is a positive measure μ on X, for which

$$a^*(f) = \int f \mathrm{d}\mu, \qquad (f \in A),$$

then it is clear that $a^*(1) = 1 = \|a^*\|$ and from Proposition 2.18 it follows that $a^* \in \langle X \rangle$.

Conversely, if $a^*(1) = 1 = \|a^*\|$, then there is an extension μ of $\mathrm{Re}a^*$ to $C_R(X)$ with the same properties: $\mu(1) = 1 = \|\mu\|$. Let

$\mu = \mu^+ - \mu^-$ be the Jordan decomposition of a real measure μ. We have

$$1 = \mu(1) = \mu^+(1) - \mu^-(1)$$

$$1 = \|\mu\| = |\mu|(1) = \mu^+(1) + \mu^-(1).$$

Therefore $\mu^-(1) = 0$, which implies $\|\mu^-\| = \mu^-(1) = 0$ and hence $\mu^- = 0$. Then μ is a positive measure with $\mu(X) = 1$ and

$$a^*(f) = \int f d\mu. \qquad (f \in A).$$

Theorem 2.20. *The Choquet boundary of X relative to A is equal to the set of all extremal points of the convex compact set* $\langle X \rangle = \{a^* \in A^* : a^*(1) = 1 = \|a^*\|\}$.

Proof. Let x be an extremal point of $\langle X \rangle$. Since X is a compact subset of A^*, it results that $x \in X$. Let $\mu \ll \varepsilon_x$ and suppose there is a compact $K \subset X - \{x\}$ with $\mu(K) > 0$. The compactness of K ensures the existence of a point $y \in K$ such that for any neighbourhood U of y, $\mu(U) > 0$. If for any neighbourhood U of y we have $\mu(U) = 1$, then for any compact F which does not contain y, $\mu(F) = 0$, hence, $\mu(X) - \{y\}) = 0$, that is, $\mu = \varepsilon_y$.

Since $\mu \ll \varepsilon_x$, $f(x) = f(y)$ for any $f \in A$, which is a contradiction because A separates X. Thus there is a neighbourhood U of y with $0 < \mu(U) < 1$. We write $r = \mu(U)$ and $\mu_1 = r^{-1}\chi_U \mu$, $\mu_2 = (1 - r)^{-1}(1 - \chi_U)\mu$. μ_1 and μ_2 are positive measures on X and $\mu_1(X) = \mu_2(X) = 1$.

If we put

$$a_1^*(f) = \int f d\mu_1 \qquad (f \in A).$$

$$a_2^*(f) = \int f d\mu_2 \qquad (f \in A)$$

then, according to Proposition 2.19 we get two functionals a_1^* and a_2^* from $\langle X \rangle$.

If $x = a_1^* = a_2^*$, then $x \in \langle K \rangle$ which is not possible since x is extremal and $K \subset X - \{x\}$.

On the other side we have

$$\mu = r\mu_1 + (1 - r)\mu_2$$

and so $x = ra_1^* + (1 - r)a_2^*$ which contradicts the extremality of x. Thus $\mu(X - \{x\}) = 0$, hence $\{x\}$ is absorbant and point (b) of Theorem 2.8 implies that x belongs to the Choquet boundary of X relative to A.

Conversely, let x be a Choquet point and suppose $x = ra_1^* + (1 - r)a_2^*$ with $0 < r < 1$ and $a_1^*, a_2^* \in \langle X \rangle$, $a_1^* \neq a_2^*$.

By Proposition 2.19 there exist positive measures μ_1, μ_2 on X, $\mu_1(X) = \mu_2(X) = 1$, such that

$$a_1^*(f) = \int f d\mu_1 \qquad (f \in A),$$

$$a_2^*(f) = \int f d\mu_2 \qquad (f \in A)$$

If we write $\mu = r\mu_1 + (1 - r)\mu_2$, then μ is a positive measure on X, $\mu(X) = 1$ and

$$f(x) = \int f d\mu \qquad (f \in A)$$

then $\mu \ll \varepsilon_x$. Since $a_1^* \neq a_2^*$, we have $\mu_1 \neq \mu_2$. Hence $\mu_1 \neq \varepsilon_x$ say, which means $\mu_1(\{x\}) < 1$. It follows that

$$\mu(\{x\}) = r\mu_1(\{x\}) + (1 - r)\,\mu_2(\{x\}) < 1$$

i.e. $\mu \neq \varepsilon_x$. Thus we have constructed a measure $\mu \ll \varepsilon_x$, different from ε_x, which contradicts the fact that x is a Choquet point.

The contradiction follows from the assumption that x is not extremal. Therefore x is extremal and the theorem is completely proved. ◇

2.6. Representing theorems

Let A be a function algebra on X, Σ its Choquet boundary and Γ its Shilov boundary. Since for any determining set S and any point $x \in X$ the conditions of Theorem 2.1 are fulfilled, we have the following.

Theorem 2.21. *For any $x \in X$ there exists a positive measure μ on X such that $\mu \ll \varepsilon_x$ and $\mu(X - \Gamma) = 0$.*

One can investigate the existence of representing measures for points in X which are supported by the Choquet boundary.

The problem is dificult in general as the Choquet boundary is an arbitrary set (sometimes even not a Borel set). In this case there is a result establishing the existence of representing measures for the points of X, which vanish on any Baire set contained in $X - \Gamma$ (Bishop — de Leeuw).

Here we treat only the case when X is metrizable.

Theorem 2.22. *Let A be a function algebra on a compact metric space X. The Choquet boundary of A is a G_δ-set.*

Proof. Let d be the metric on X and V_n the set of all points $x \in X$ for which there is a function $f \in A$ such that

(i) $\|f\| \leqslant 1$

(ii) $|f(x)| > 3/4$

(iii) $|f(y)| < 1/4$ for any $y \in X$ with $d(x, y) > 1/n$.

V_n are obviously open sets and $\Sigma = \bigcap V_n$. Indeed, since any Choquet point is, in the metrizable case, a peak point, we have $\Sigma \subset \bigcap V_n$. The converse inclusion results from point (d) of Theorem 2.8 ◇

Theorem 2.23 *(Choquet). Let A be a function algebra on a compact metric space X and Σ its Choquet boundary.*

For any $x \in X$ there is a measure $\mu \ll \varepsilon_x$ such that $\mu(X - \Sigma) = 0$.

Proof. Suppose X is embedded in A^* as in §2.5. Clearly we may assume $X = \langle X \rangle$. For a function $w \in C_R(X)$ we define

$$\overline{w}(x) = Q_x^X(w) = \inf \{u(x): u \in \mathrm{Re}A,\ u \geqslant w\}.$$

$\overline{w}$ is upper semicontinuous and $w \leqslant \overline{w}$, $w = \overline{w}$ if $w \in \mathrm{Re}A$.

Since X is metrizable, ReA is separable. Let $u_n \in \mathrm{Re}A$ be such that $\|u_n\| = 1$ and $\{u_n\}$ be a dense set in the unit sphere of ReA. Let $w = \Sigma\, 2^{-n} u_n^2$, $w \in C_R(X)$. Let B denote $\mathrm{Re}A + Rw$ the subspace generated by ReA and w in $C_R(X)$. Let us fix an $x_0 \in X$. Since the functional $Q_{x_0}^X$ is a seminorm, by the Hahn-Banach theorem there is a functional μ on $C_R(X)$ such that $\mu(w) \leqslant Q_{x_0}^X(w)$ and $\mu(u + rw) = u(x_0) + rQ_{x_0}^X(w)$. If $g \in C_R(X)$ and $g \leqslant 0$, then $0 \geqslant Q_{x_0}^X(g) \geqslant \mu(g)$, and thus μ is a positive measure. Obviously $\mu \ll \varepsilon_{x_0}$.

Since $w \leqslant \bar{w}$, we have $\mu(w) \leqslant \mu(\bar{w})$. On the other hand, for $u \in \operatorname{Re} A$, $u \geqslant w$ and so $u \geqslant \bar{w}$. Therefore,

$$u(x_0) = \mu(u) \geqslant \mu(\bar{w}).$$

Then

$$\mu(\bar{w}) \leqslant \bar{w}(x_0) = Q_{x_0}^X(w) = \mu(w)$$

and $\mu(\bar{w}) = \mu(w)$.

But $w \leqslant \bar{w}$, so that μ vanishes outside the set $\{x \in X: w(x) = \bar{w}(x)\}$.

On the other hand, this set is contained in Σ. Indeed, let x be such that $w(x) = \bar{w}(x)$ and $x = \frac{1}{2} z + \frac{1}{2} y$ with $z \neq y$ in X. $\operatorname{Re} A$ separates X and since $\{u_n\}$ is dense in the unit sphere of $\operatorname{Re} A$, there is an n such that $u_n(y) \neq u_n(z)$.

Then

$$u_n^2\left[\frac{1}{2} z + \frac{1}{2} y\right] = \frac{1}{4} u_n^2(z) + \frac{1}{4} u_n^2(y) + \frac{1}{2} u_n(z) u_n(y) <$$

$$< \frac{1}{2} u_n^2(z) + \frac{1}{2} u_n^2(y)$$

and, therefore,

$$w\left(\frac{1}{2} z + \frac{1}{2} y\right) < \frac{1}{2} w(z) + \frac{1}{2} w(y).$$

Hence

$$w(x) < \frac{1}{2} w(z) + \frac{1}{2} w(y) \leqslant \frac{1}{2} \bar{w}(z) + \frac{1}{2} \bar{w}(y) \leqslant \bar{w}(x)$$

since

$$\bar{w}\left(\frac{1}{2} z + \frac{1}{2} y\right) = \inf\left\{u\left(\frac{1}{2} z + \frac{1}{2} y\right): u \in \operatorname{Re} A,\ u \geqslant w\right\} \geqslant$$

$$\geqslant \frac{1}{2} \inf\{u(z), u \in \operatorname{Re} A, u \geqslant w\} + \frac{1}{2} \inf\{u(y): u \in \operatorname{Re} A, u \geqslant w\} =$$

$$= \frac{1}{2} \bar{w}(z) + \frac{1}{2} \bar{w}(y).$$

We have arrived at a contradiction, which shows that x is extremal and therefore $x \in \Sigma$.

The theorem is completely proved. ◇

Notes

This chapter presents, in the frame of function algebras, different results of integral representations on minimal boundaries; the origin of these results can be found in contexts of: commutative Banach algebras, convexity theory in locally convex vector spaces, potential theory, convex cones, etc.

E. Bishop [1] was the first to remark the connection between the theory of minimal boundaries for function algebras and the Choquet theory.

Theorem 2.1 and its corollaries have been exposed by many authors in different contexts (cf. R. Arens, I. M. Singer [1], H. Bauer [1], G. E. Shilov [1], etc.) The form presented here was given in a more general frame in N. Boboc, A. Cornea [1].

Theorem 2.8 contains several characterizations of the points of the Choquet boundary, which appear in different mathematical fields in works as: H. Bauer [1], E. Bishop [1], E. Bishop, K. de Leeuw [1], N. Boboc, A. Cornea [1], D. A. Edwards [1], R. Phelps [1].

The characterization of the Shilov boundary points presented in Theorem 2.16 appeared for the first time in G. E. Shilov [1]. Theorem 2.21 belongs to G. E. Shilov [1] and Theorem 2.23 derives from the well-known classical Choquet theorem (cf. G. Choquet [1]).

The present chapter has been written following mainly N. Boboc, A. Cornea [1], K. Hoffman [3] and R. Phelps [1].

CHAPTER 3

Algebras on the maximal ideal space

3.1. The maximal ideal space

Let A be a function algebra on X. A is a commutative Banach algebra with unit element $e = 1$. Moreover

$$\|f^2\| = \|f\|^2 \qquad (f \in A). \tag{3.1.1}$$

Let $\mathfrak{M}$ be the maximal ideal space of A. We have already seen in paragraph 2.5 that the map $x \to e_x$, $e_x \in A^*$, defined by

$$e_x(f) = f(x) \qquad (f \in A)$$

is a homeomorphic embedding of X into A^*, A^* being endowed with the weak*-topology. Since e_x are obviously multiplicative, by this embedding X is a compact subset of $\mathfrak{M}$.

From (3.1.1), the homomorphism $f \to \hat{f}$ of A into $C(\mathfrak{M})$ is an isometric isomorphism and because of this A will be considered as a subalgebra of $C(\mathfrak{M})$. Of course, A is a function algebra on $\mathfrak{M}$. (3.1.1) also shows that X is a determining set of $\mathfrak{M}$.

Proposition 3.1. *The following assertions are equivalent:*

(a) $X = \mathfrak{M}$,

(b) *for any finite system $f_1, \ldots, f_n$ of functions of A, which have no common zeros on X, there exists a system $g_1, \ldots, g_n$, of functions of A, such that $\Sigma f_i g_i = 1$.*

Proof. (a) → (b). Let I be the ideal generated by $f_1,\dots,f_n$.

If $I \neq A$ then there is a maximal ideal $M = \operatorname{Ker}\varphi$, $\varphi \in \mathfrak{M}$ such that $I \subset M$. Therefore, $\varphi(f_i) = 0$, $i = 1, 2,\dots, n$, which is not possible. Thus $I = A$ and there exist $g_1,\dots, g_n$ in A such that

$$\sum f_i g_i = 1.$$

(b) → (a). Let $\varphi \in \mathfrak{M}$ and suppose $\varphi \notin X$. Since X is a compact subset of $\mathfrak{M}$, there is a neighbourhood $U_{\varepsilon; f_1,\dots,f_n}(\varphi)$ of φ such that $U_{\varepsilon; f_1,\dots,f_n}(\varphi) \cap X = \emptyset$. We can suppose $f_1(\varphi) = \dots = f_n(\varphi) = 0$, so that it is clear that $f_1,\dots,f_n$ have no common zeros on X. (b) asserts that there are $g_1,\dots, g_n$ in A with $\Sigma f_i g_i = 1$ which is impossible, and that completes the proof.

It is also worth noting that since X is a determining set of $\mathfrak{M}$, the Choquet (Shilov) boundary of $\mathfrak{M}$ relative to A is identical to the Choquet (Shilov) boundary of X relative to A. Also, for any $\varphi \in \mathfrak{M}$ there exists a representing measure for φ with support contained in the Shilov boundary.

3.2. Locally analytic functions

From now until the end of the chapter we shall suppose that A is function algebra on the maximal ideal space $\mathfrak{M}$.

We call the *common spectrum* of the elements $f_1,\dots, f_n \in A$ the set in C^n defined by

$$\sigma(f_1, f_2,\dots, f_n) = \{z \in C^n : z = (f_1(\varphi),\dots, f_n(\varphi)),\ \varphi \in \mathfrak{M}\}.$$

By $\hat{\sigma}(f_1,\dots, f_n)$ we denote the convex polynomial hull of $\sigma(f_1,\dots, f_n)$. Both $\sigma(f_1,\dots, f_n)$ and $\hat{\sigma}(f_1,\dots, f_n)$ are compact subsets of C^n.

Lemma 3.2. *Let A be a function algebra on $\mathfrak{M}$. Let $f_1,\dots,f_n \in A$ and U be a neighbourhood of $\sigma(f_1,\dots, f_n)$ in C^n. Then there exist $f_{n+1},\dots,f_m \in A$ such that $\pi[\hat{\sigma}(f_1,\dots, f_m)] \subset \hat{\sigma}(f_1,\dots, f_n) \cap U$, where $\pi: C^m \to C^n$ is the natural projection of C^m onto C^n.*

Proof. Let $\sigma = \sigma(f_1,.,f_n)$. If $\hat{\sigma} \subset U$ nothing remains to be proved. Thus let $z^0 \in \hat{\sigma} - U$, $z^0 = (z_1^0,\dots, z_n^0)$. The functions $f_1 - z_1^0$, $f_2 - z_2^0,\dots, f_n - z_n^0$ then have no common zeros on $\mathfrak{M}$. Proposition 3.1 asserts the existence of $g_1,\dots, g_n \in A$ with the property $\Sigma g_i(f_i - z_i^0) = 1$.

Let $\sigma' = \sigma(f_1,\dots, f_n, g_1,\dots, g_n)$ and $\pi\colon C^{2n} \to C^n$ be the natural projection. Clearly $\pi(\sigma') \subset \sigma$ and hence $\pi(\hat{\sigma}') \subset \hat{\sigma}$. Since $\Sigma z_{n+i}(z_i - z_i^0) = 1$ on σ', the same is true on $\hat{\sigma}'$. We have

$$\{z \in C^n : \sum z_{n+1}(z_i - z_i^0) = 1, z_i = z_i^0, 1 \leqslant i \leqslant n\} = \varnothing, \text{ thus}$$

$$z^0 \notin \pi(\hat{\sigma}').$$

But $\hat{\sigma}'$ is compact so that there is a neighbourhood U_{z^0} of z^0 such that $\pi(\hat{\sigma}') \cap U_{z^0} = \varnothing$.

Hence, $\hat{\sigma} - U$, as a compact set can be covered with a finite number of neighbourhoods $U_1,\dots, U_k$ constructed in the same way as above. Let $f_{n+1},\dots, f_m$ be the finite system of functions which appears when constructing $U_1,\dots, U_k$. Then

$$\pi[\hat{\sigma}(f_1,\dots, f_n, f_{n+1},\dots, f_m)] \subset \hat{\sigma}(f_1,\dots, f_n) \cap U. \diamond$$

Theorem 3.3. *(Shilov-Arens-Calderon): Let A be a function algebra on $\mathfrak{M}$-the maximal ideal space, and $f_1,\dots, f_n \in A$. If F is a holomorphic function on a neighbourhood U of the common spectrum $\sigma(f_1,\dots,f_n)$ of $(f_1,\dots, f_n)$, then $F(f_1, f_2,\dots, f_n) \in A$.*

Proof. Lemma 3.2 asserts the existence of $f_{n+1},\dots, f_m \in A$ for which $\pi[\hat{\sigma}(f_1,\dots, f_m)] \subset \hat{\sigma} \cap U$. Then $F \circ \pi$ is holomorphic on a neighbourhood of the compact polynomially convex set $\hat{\sigma}(f_1,\dots,f_m)$. From Oka's theorem (Proposition 1.6) we find that there is a sequence of polynomials $P_k(z_1,\dots, z_m)$ uniformly convergent to $F \circ \pi$ on $\hat{\sigma}(f_1,\dots,f_m)$. Hence $P_k(f_1,\dots, f_m)$ converges uniformly to $F(f_1,\dots, f_n)$ on $\hat{\sigma}$ and therefore $F(f_1,\dots, f_n) \in A$. $\diamond$

Theorem 3.4. *(Shilov). Let A be a function algebra on $\mathfrak{M}$-the maximal ideal space. Let K be a closed and open set in $\mathfrak{M}$. Then $\chi_k \in A$.*

Proof. Let $x \in K$ and $y \notin K$. Since A separates the points of $\mathfrak{M}$ there is an $f \in A$ such that $f(x) = 1$, $f(y) = 0$.

Let

$$U_f = \left\{\varphi \in \mathfrak{M} : |f(\varphi) - 1| < \frac{1}{4}\right\}, \quad V_f = \left\{\varphi \in \mathfrak{M} : |f(\varphi)| < \frac{1}{4}\right\}.$$

$\mathfrak{M} - K$ is compact and, therefore, it can be covered with a finite number of neighbourhoods $V_{f_1}, \ldots, V_{f_n}$ constructed as above. Let $U_{f_1}, \ldots, U_{f_n}$ be their corresponding neighbourhoods and $U_x = \bigcap_{i=1}^{n} U_{f_i}$. Hence we have constructed for any $x \in K$ a neighbourhood U_x, the functions $f_1, \ldots, f_n$ and the open sets $V_{f_1}, \ldots, V_{f_n}$, such that

$$V_{f_1}, \ldots, V_{f_n} \text{ cover } \mathfrak{M} - K, \ U_x \cap V_{f_i} = \varnothing, i = 1, \ldots, n,$$

$$|f_i| < \frac{1}{4} \text{ on } V_{f_i}, \ |f_i - 1| < \frac{1}{4} \text{ on } U_x.$$

Now we cover K with a finite number of open sets $U_1, \ldots, U_n$ constructed as above, and let $f_{11}, \ldots, f_{1n_1}, f_{21}, \ldots, f_{2n_2}, \ldots, f_{m_1}, \ldots, f_{mn_m}$ be the finite system of functions which results from the construction. Let us number these $f_1, \ldots, f_p$.

Let

$$\sigma_k = \{z \in C^p : z = (f_1(x), \ldots, f_p(x)), \ x \in K\}$$

$$\sigma_k' = \{z \in C^p : z = (f_1(x), \ldots, f_p(x)), \ x \in \mathfrak{M} - K\}.$$

The properties of $f_1, \ldots, f_p$ give $\sigma_k \cap \sigma_k' = \varnothing$ and, therefore, obviously, $\sigma(f_1, \ldots, f_p) \subset \sigma_k \cup \sigma_k'$. Let U and V be two open sets in C^p such that $U \cap V = \varnothing$ and $\sigma_k \subset U$, $\sigma_k' \subset V$. The function F defined by

$$F(z) = \begin{cases} 1 & \text{for } z \in U \\ 0 & \text{for } z \in V \end{cases}$$

is clearly holomorphic on $U \cup V$, i.e. on a neighbourhood of $\sigma(f_1, \ldots, f_p)$. Then from Theorem 3.3 it results that $\chi_k = F(f_1, \ldots, f_p)$ belongs to A.

The theorem is completely proved. ◇

Corollary 3.5. *If $\mathfrak{M}$ is totally nonconnected then $A = C(\mathfrak{M})$.*

Corollary 3.6. *If x is an isolated point of $\mathfrak{M}$ then it is a peak point for A.*

3.3. The local maximum modulus principle

First we prove the following lemma on functions of several complex variables.

Lemma 3.7. *Let K be a compact in C^n, $P \subset K$ a closed set and U a neighbourhood of P. Let h be holomorphic in U, $h = 0$ on P and $\mathrm{Re}\, h < 0$ on $U - P$. Let f be holomorphic on a neighbourhood V of $K (U \subset V)$, invertible on $K - U$ and $f = he^{gh}$ in U, with g holomorphic on U. Then there exists a holomorphic function F in a neighbourhood of K such that $F = 1$ on P and $|F| < 1$ on $K - P$.*

Proof. Since $f = he^{gh}$ on U, we have

$$h = fe^{-gh} = f + g'f^2$$

with g' a holomorphic function on U. It follows that, on $U \cap (K - P)$

$$0 > \mathrm{Re}\, h = \mathrm{Re} f + \mathrm{Re} f^2 g' \geqslant \mathrm{Re} f - |f|^2 |g'| \geqslant \mathrm{Re} f - M|f|^2$$

where $M = \sup\limits_U |g'|$. Then, on $U \cap (K - P)$

$$|f|^2 > \frac{\mathrm{Re} f}{M}$$

and, therefore,

$$\left| f - \frac{1}{2M} \right| = \sqrt{\left(\mathrm{Re} f - \frac{1}{2M}\right)^2 + (\mathrm{Im} f)^2} =$$

$$= \sqrt{(\mathrm{Re} f)^2 + (\mathrm{Im} f)^2 + \left(\frac{1}{2M}\right)^2 - \frac{\mathrm{Re} f}{M}} > \sqrt{|f|^2 + \left(\frac{1}{2M}\right)^2 - |f|^2} = \frac{1}{2M}$$

But f does not vanish on $K - U$, and hence there exists an $\varepsilon > 0$ such that

$$|f - \varepsilon| > \varepsilon.$$

This implies the existence of an $\varepsilon > 0$ such that

$$|f - \varepsilon| > \varepsilon$$

on $K - P$.
Since $f = 0$ on P, $(f - \varepsilon)^{-1}$ exists and is holomorphic on a neighbourhood of K. Let

$$F = -\varepsilon(f - \varepsilon)^{-1}.$$

F is analytic on a neighbourhood of K, $F = 1$ on P and $|F| < 1$ on $K - P$. The proof is complete. ◇

Let A be a function algebra on $\mathfrak{M}$-the maximal ideal space. A closed subset F of $\mathfrak{M}$ is called a *locally peak set* if there exist a neighbourhood U of F and a function $f \in A$ such that $f = 1$ on F and $|f| < 1$ on $U - F$.

Theorem 3.8. *(H. Rossi). Any locally peak set is a peak set.*

Proof. Let F be a locally peak set, i.e. there exist a neighbourhood U of F and an $f \in A$, with $f = 1$ on F and $|f| < 1$ on $U - F$. Let $g_0 = f - 1$. Then $g_0 = 0$ on F and Re $g_0 < 0$ on $U - F$. Let $x \in F$ and V be one of its neighbourhoods of the form

$$V = \{\varphi \in \mathfrak{M} : |g_i(\varphi)| < 2,\ 1 \leqslant i \leqslant n,\ g_i \in A,\ g_i(x) = 0\}$$

such that $x \in V \subset U$.

$$\text{Let } V_x = \{\varphi \in \mathfrak{M} : |g_i(\varphi)| < 1,\ 1 \leqslant i \leqslant n\}.$$

We have $x \in V_x \subset V$ and since F is compact, it can be covered with a finite number $V_1, \ldots, V_p$ of such neighbourhoods:

$$V_j = \{\varphi \in \mathfrak{M} : |g_i(\varphi)| < 1,\ t_{j-1} \leqslant i \leqslant t_j\},\ j = 1, \ldots, p.$$

$$\text{Let } t = t_p,\quad M = \max_{1 \leqslant i \leqslant t} \|g_i\|,\quad \sigma = \sigma(g_0, g_1, \ldots, g_t).$$

Denote by $z = (z_0, z_1, \ldots, z_t)$ the points of C^{t+1}; we define the continuous function $G\colon \mathfrak{M} \to C^{t+1}$ by

$$\varphi \to (g_0(\varphi), g_1(\varphi), \ldots, g_t(\varphi)).$$

Let
$\Delta = \{z \in C^{t+1}\colon$ there exists a j such that $|z_i| < 1$ for $t_{j-1} \leqslant i \leqslant t_j$, and $|z_i| < M$ for other $i\}$.

One can easily verify that:

1) $\Delta \cap \sigma \subset \{z \in C^{t+1}\colon \operatorname{Re} z_0 \leqslant 0\}$

2) $\Delta \cap \sigma \cap \{z \in C^{t+1}\colon \operatorname{Re} z_0 = 0\} = \Delta \cap \sigma \cap \{z \in C^{t+1}\colon z_0 = 0\}$

3) $G(F) \subset \Delta \cap \sigma \cap \{z \in C^{t+1}\colon z_0 = 0\}$

4) $G^{-1}(\sigma \cap \bar{\Delta}) \subset U$

The compact set $(\sigma \cap \bar{\Delta}) - \Delta$ is included in $\{z \in C^{t+1}\colon \operatorname{Re} z_0 < 0\}$. Indeed, if $z \in \sigma \cap \bar{\Delta}$, then from 4) there exists an $x \in U$ such that $z = G(x)$, hence $\operatorname{Re} z_0 = \operatorname{Re} g_0(x) \leqslant 0$. If $\operatorname{Re} g_0(x) = 0$, then $x \in F$ and from 3) we get $x \in \Delta$.

Let D_1 be a neighbourhood of $\sigma - \Delta$ and D_2 a neighbourhood of $(\sigma \cap \bar{\Delta}) - \Delta$, such that $D_2 \subset \{z \in C^{t+1}\colon \operatorname{Re} z_0 < 0\}$ and $D_0 = (D_1 - \bar{\Delta}) \cup D_2$. If $z \in \sigma - \Delta$ and $z \in \bar{\Delta}$, then $z \in (\sigma \cap \bar{\Delta}) - \Delta \subset D_2$, and therefore $\sigma - \Delta \subset D_0$. Moreover $D_0 \cap \Delta = D_2 \cap \Delta \subset \{z \in C^{t+1}\colon \operatorname{Re} z_0 < 0\}$. Let $D = \Delta \cup D_0$.

Since D is a neighbourhood of σ, from Lemma 3.2 we can suppose $\hat{\sigma} \subset D$ and follow the same construction in a C^n with $n \geqslant t+1$.

There exists a polynomial polyhedron Q such that $\hat{\sigma} \subset Q \subset D$.

Indeed, let $r > 0$ be such that $\hat{\sigma} \subset \Delta(0, r)$. Let V be a neighbourhood of $\hat{\sigma}$ with $\bar{V} \subset D$ and $\bar{V}$ compact.

For any $p \in \partial V$ (the topological boundary) there is a polynomial P such that

$$|P(p)| > \max_{z \in \hat{\sigma}} |P(z)|$$

since $\hat{\sigma}$ is polynomially convex. By multiplication with a constant we can suppose

$$|P(p)| > r > \max_{z \in \hat{\sigma}} |P(z)|.$$

Let

$$V_p = \{z \in C^n : |P(z)| > r\}.$$

We now cover ∂U with a finite number of such open sets and let $P_1, P_2, \ldots, P_m$ be the corresponding polynomials. The polynomial polyhedron $Q = W(P_1, \ldots, P_m, r)$ satisfies $\hat{\sigma} \subset Q \subset D$.

Let $Q_0 = D_0 \cap Q$, $Q_1 = \Delta \cap Q$.

Q_0, Q_1 are open sets and $Q = Q_0 \cup Q_1$, $Q_0 \cap Q_1 \subset D_0 \cap \Delta \subset \{z \in \in C^n : \operatorname{Re} z_0 < 0\}$. Then, $\log z_0$ is a holomorphic function on $Q_0 \cap Q_1$ and by putting $h_{12} = - h_{21} = z_0^{-1} \log z_0$ we obtain on Q the system of Cousin data $\{Q_i, h_{ij}\}$.

Cousin's theorem (Prop. 15) asserts the existence of holomorphic functions g_0, g_1, on Q_0, Q_1, respectively, for which $g_0 - g_1 = z_0^{-1} \log z_0$ on $Q_0 \cap Q_1$.

Let

$$f_1(z) = \begin{cases} e^{g_0 z_0} & z \in Q_0 \\ z_0 e^{g_1 z_0} & z \in Q_1. \end{cases}$$

Clearly f_1 is holomorphic on Q.

If in the preceding lemma we put

$$K = \hat{\sigma},\ P = \hat{\sigma} \cap \Delta \cap \{z \in C^n : z_0 = 0\},\ h = z_0,\ U = Q_1,\ f = f_1,$$

since all conditions are easily verified, there exists a holomorphic function F in the neighbourhood of $\hat{\sigma}$ such that

$$F(z) = 1 \text{ for } z \in \hat{\sigma} \cap \Delta \cap \{z \in C^n : z_0 = 0\} \text{ and } |F(z)| < 1$$

for the other points of $\hat{\sigma}$. From Shilov-Calderon's theorem (Theorem 3.3) there follows the existence of a function $f \in A$ with $f(\varphi) = F(G(\varphi))$. Then $f(\varphi) = 1$ for $\varphi \in F$ and $|f(\varphi)| < 1$ for $\varphi \notin F$. Therefore F is a peak set and the theorem is completely proved. ◇

Corollary 3.9. *For any $f \in A$ and $t \geqslant 0$, the intersection of each connected component of the set $\{\varphi \in \mathfrak{M} : |f(\varphi)| \geqslant t\}$ with the Shilov boundary is nonvoid.*

Corollary 3.10. *If V is an open subset of $\mathfrak{M}$ with $V \subset \mathfrak{M} - \Gamma$, then the maximum of the modulus of any function from A on V is attained on the topological boundary of V.*

Corollary 3.11. *The functions of A can not attain strictly local maxima in $\mathfrak{M} - \Gamma$.*

3.4. Gleason parts

Let A be a function algebra on X, $\mathfrak{M}$ the maximal ideal space and Γ its Shilov boundary.

Theorem 3.12. *Let $\varphi_1, \varphi_2 \in \mathfrak{M}$. The following assertions are equivalent:*

(i) $\|\varphi_1 - \varphi_2\| < 2$.

(ii) *There exists a $C < 1$ such that for any $f \in A$ with $\|f\| < 1$ and $f(\varphi_1) = 0$ we have $|f(\varphi_2)| \leqslant C$.*

(iii) *For any sequence h_n in A, with $\|h_n\| \leqslant 1$ and $\lim |h_n(\varphi_2)| = 1$ we have $\lim |h_n(\varphi_1)| = 1$.*

(iv) *There exists an $a > 0$ such that for any strictly positive function $u \in \operatorname{Re} A$*

$$\frac{1}{a} \leqslant u(\varphi_1)/u(\varphi_2) \leqslant a.$$

(v) *There exists a $c < 1$ and the representing measures μ_1, μ_2 for φ_1, φ_2 respectively, with their support in Γ such that $c\mu_1 \leqslant \mu_2$, $c\mu_2 \leqslant \mu_1$.*

Proof. Suppose (ii) is not true. For $\varepsilon > 0$ we define the function F on the unit disc, by

$$F(z) = \frac{z - (1 - \varepsilon)}{1 - (1 - \varepsilon)z} \qquad (|z| \leqslant 1).$$

For sufficiently small ε, F is analytic on the closed unit disc, $|F(z)| \leqslant 1$ for any z with $|z| \leqslant 1$, $F(1) = 1$ and $F(0) = -1 + \varepsilon$. Hence, there is an $\eta > 0$ such that for $|z| \leqslant 1$, $|1 - z| \leqslant \eta$, $|1 - F(z)| < 1$. Since (ii) is not true, there is an $f \in A$ with $\|f\| \leqslant 1$ for which $f(\varphi_1) = 0$ and $f(\varphi_2) = 1 - \eta$. Let $g \in A$ be defined by $g(x) = F(f(x))$ for $x \in X$. Then $\|g\| \leqslant 1$ and

$$|g(\varphi_1) - g(\varphi_2)| = |-1 + \varepsilon - F(1 - \eta)| = |-2 + \varepsilon - (F(1 - \eta) - \\ -1)| \geqslant 2 - \varepsilon - |1 - F(1 - \eta)| \geqslant 2 - 2\varepsilon.$$

Therefore

$$\|\varphi_1 - \varphi_2\| = \sup \{|g(\varphi_1) - g(\varphi_2)|;\, g \in A,\ \|g\| \leqslant 1\} = 2,$$

i.e. (i) is *false*.

Suppose now (iii) is not true. Then there is a $C < 1$ such that for any $\varepsilon > 0$ we can find an

$$f \in A \text{ with } \|f\| \leqslant 1,\ |f(\varphi_1)| \leqslant C,\ |f(\varphi_2)| > 1 - \varepsilon.$$

For a fixed $\varepsilon_1 > 0$ there exists an $\eta_1 > 0$ and an $m > 0$ such that

$$1 - \varepsilon_1 \leqslant \frac{|z - z_0|}{m} \leqslant 1$$

holds for any $|z_0| \leqslant C$ and any $|z| \leqslant 1$, $|1 - z| < \eta_1$.

For any $\varepsilon_2 > 0$ we can find a function F, continuous on the closed unit disc and analytic on its interior (i.e. in the standard algebra), such that $\|F\| \leqslant 1$, $F(1) = 1$ and $|F(z)| < \varepsilon_2$ for any $|z| \leqslant 1$, $|1 - z| \geqslant \eta_1$. Such a function indeed exists since $z = 1$ is a Choquet point for the standard algebra.

Then, there is an η_2, $0 < \eta_2 < \eta_1$ with $|1 - F(z)| < \varepsilon$ for $|z| \leqslant 1$, $|1 - z| \leqslant \eta_2$. Since (iii) is not true, there is a function $f \in A$ for which $\|f\| \leqslant 1$, $|f(\varphi_1)| \leqslant C$ and $f(\varphi_2) = 1 - \eta_2$. Let $z_0 = f(\varphi_1)$ and G the function in the standard algebra defined by

$$G = \frac{z - z_0}{m} F.$$

If we choose a sufficiently small ε_2, $\|G\| \leqslant 1$; we have also

$$G(z_0) = 0 \text{ and } |G(z)| > (1 - \varepsilon_1)^2 \text{ for } |z| \leqslant 1,\ |1 - z| \leqslant \eta_2.$$

Let $g = G(f)$. Then one easily verifies that $g \in A$,

$$\|g\| \leqslant 1,\ g(\varphi_1) = 0 \text{ and } |g(\varphi_2)| > 1 - 2\varepsilon_1 + \varepsilon_1^2$$

which contradicts (ii). Therefore (ii) → (iii) is proved.

The assertion (i) is symmetric in φ_1, φ_2, hence the same is true for (ii) and (iii).

Suppose (iii) true and let $0 < c < 1$ be such that there exists a strictly positive function u in ReA with $u(\varphi_2) = 1$, and $u(\varphi_1) < c$. Now let $v \in \text{Re}A$ with $u + iv \in A$ and $f = e^{-(u+iv)}$.

$$\text{Then } \|f\| \leqslant 1, |f(\varphi_1)| > e^{-c} \text{ and } |\ \varphi_2)| = e^{-1}.$$

Since we can take c as small as we like, then, obviously, there is a sequence (f_n) of functions in A such that $\|f_n\| \leqslant 1$, $f_n(\varphi_1) \to 1$ and $|f_n(\varphi_2)| = e^{-1}$ which contradicts (iii).

Therefore, there exists a $0 < c < 1$ with the property that for any strictly positive function u in ReA we have

$$u(\varphi_1) \geqslant cu(\varphi_2).$$

From symmetry there follows that there exists a c with

$$u(\varphi_2) \geqslant cu(\varphi_1),$$

hence there exists an $a < 1$ such that

$$\frac{1}{a} \leqslant \frac{u(\varphi_1)}{u(\varphi_2)} \leqslant a$$

i.e. (iv) is true, and thus (iii) → (iv) is proved.

Suppose now that (iv) is true. Then there is a $c = a^{-1}$, $0 < c < 1$, such that

$$u(\varphi_1) - cu(\varphi_2) \geqslant 0$$

$$u(\varphi_2) - cu(\varphi_1) \geqslant 0$$

for any positive u in ReA. We define on A the linear functionals

$$L_1(u) = (1 - c)^{-1}\,[u(\varphi_1) - cu(\varphi_2)]$$

$$L_2(u) = (1 - c)^{-1}\,[u(\varphi_2) - cu(\varphi_1)]$$

which are obviously positive, $L_1(1) = L_2(1) = 1$ and therefore $\|L_1\| = \|L_2\| = 1$. Let m_1, respectively m_2 be extensions of L_1, respectively L_2 to $C(\Gamma)$, with the same norm. Since $m_i(1) = \|m_i\|$, $i = 1, 2$, the measures m_i are positive and

$$f(\varphi_1) - cf(\varphi_2) = (1 - c) \int f dm_1,$$

$$f(\varphi_2) - cf(\varphi_1) = (1 - c) \int f dm_2.$$

According to these relations we get

$$f(\varphi_1) = cf(\varphi_2) + (1 - c) \int f dm_1 = c^2 f(\varphi_1) + c(1 - c) \int f dm_2 + (1 - c) \int f dm_1$$

$$f(\varphi_2) = c^2 f(\varphi_2) + c(1 - c) \int f dm_1 + (1 - c) \int f dm_2$$

that is

$$f(\varphi_1) = (1 + c)^{-1}\,[c \int f dm_2 + \int f dm_1]$$

$$f(\varphi_2) = (1 + c)^{-1}\,[c \int f dm_1 + \int f dm_2].$$

If we write

$$\mu_1 = (1 + c)^{-1} (cm_2 + m_1)$$

$$\mu_2 = (1 + c)^{-1} (cm_1 + m_2),$$

then the measures μ_1, μ_2 are representing measures for φ_1 and φ_2, respectively, with the support in Γ and we have

$$c\mu_1 \leqslant \mu_2, \quad c\mu_2 \leqslant \mu_1.$$

This proves (iv) → (v).

The implication (v) → (iv) is obviously true. It remains to prove that (iv) → (i). Let (i) be not true, that is, $\|\varphi_1 - \varphi_2\| = 2$. Then it is clear that for any $\varepsilon > 0$ there is an $f \in A$, $\|f\| \leqslant 1$ such that $f = u + iv$ implies $u(\varphi_1) < -1 + \varepsilon$, $u(\varphi_2) > 1 - \varepsilon$. If we write $u_0 = u + 1$, u_0 is a strictly positive function in ReA and

$$\frac{u_0(\varphi_1)}{u_0(\varphi_2)} < \frac{\varepsilon}{2 - \varepsilon}$$

which contradicts (iv).

The proof is thus complete. ◇

The points $\varphi_1, \varphi_2 \in \mathfrak{M}$ are said to be *Gleason equivalent* if they satisfy one of the (equivalent) assertions of Theorem 3.12. As follows from 3.12, this relation is an equivalence relation. An equivalence class relative to it is called a *Gleason part* of $\mathfrak{M}$.

The principal part of Theorem 3.12 is the following theorem of E. Bishop.

Theorem 3.13. *Let $\varphi_1, \varphi_2 \in \mathfrak{M}$ and μ_1 be a representing measure for φ_1 with its support in Γ. If φ_1 and φ_2 belong to the same Gleason part, then there is a representing measure μ_2 for φ_2 with support in Γ and a constant c, $0 < c < 1$ such that $c\mu_1 \leqslant \mu_2$.*

Proof. Since φ_1 and φ_2 belong to the same Gleason part, there exist two representing measures m_1, m_2 for φ_1, φ_2 respectively, with supports in Γ, and a constant $0 < c < 1$ such that $cm_1 \leqslant m_2$. If we put

$$\mu_2 = (m_2 - cm_1) + c\mu_1$$

then, of course, μ_2 is a representing measure for φ_2, with support in Γ and $c\mu_1 \leqslant \mu_2$.

Corollary 3.14. *The points of the Choquet boundary of A form point Gleason parts.*

To close this chapter we prove the following theorem on representing measures.

Theorem 3.15. *Let A be a function algebra on X and $\mathfrak{M}$ the space of its maximal ideals. Let $\varphi \in \mathfrak{M}$ and ν be a complex measure on X such that*

$$\varphi(f) = \int f d\nu \qquad (f \in A).$$

Then there exists a representing measure μ for φ with support in X which is absolutely continuous with respect to ν.

Proof. Let H be the Hilbert space obtained by closing A in $L^2(\mathrm{d}|\nu|)$ and H_0 be its subspace generated by all the elements of A which vanish at φ. Since for any $f \in A$ with $\varphi(f) = 0$ we have

$$\int |1-f|^2 \mathrm{d}\,|\nu| \geqslant |\int (1-f)^2 \,\mathrm{d}\,\nu| = 1,$$

it follows that $H_0 \neq H$. Let $h \in H$ orthogonal on H_0 such that

$$\int |h|^2 \mathrm{d}|\nu| = 1$$

and $d\mu = |h|^2 \mathrm{d}|\nu|$. For any $f \in A$, $\varphi(f) = 0$ we have $fh \in H_0$ and so

$$\int f \mathrm{d}\mu = \int |h|^2 f \mathrm{d}|\nu| = \int f h \bar{h} \mathrm{d}|\nu| = 0.$$

Since μ is a positive measure and $\mu(X) = 1$, μ is a representing measure for φ.

μ is obviously absolutely continuous with respect to ν and therefore the proof is complete. ◇

Notes

The maximal ideal space is one of the basic elements of Gelfand theory of commutative Banach algebras. The study of the structure of this space imposed itself also in the theory of function algebras through the deep results due to mathematicians such as R. Arens and A. Calderon, H. S. Bear, E. Bishop, E. A. Gleason, H. Hoffman, H. Rossi, G. E. Shilov, J. Wermer, etc.

Theorem 3.3 has been enounced by G. E. Shilov [3] and proved by R. Arens and A. Calderon [1]. Proof of Theorem 3.4 is given in G. E. Shilov [2] and Theorem 3.8 is proved by H. Rossi [1]. The present version follows mainly E. Gunning, H. Rossi [1].

The equivalence relation which defines the Gleason parts has been introduced by A. Gleason [1]. Other characterizations of this equivalence relation, which can be found in Theorem 3.12, are due to H. S. Bear [2] and E. Bishop [4]. Theorem 3.15 belongs to K. Hoffman [3].

CHAPTER 4

Approximation and Interpolation

4.1. Restrictions

Let A be a function algebra on X and F a closed subset of X. Let $A|F$ be the algebra of continuous functions on F which are restrictions to F of functions in A. Denote by kF the closed ideal of A defined by

$$kF = \{f \in A : f = 0 \text{ on } F\}.$$

Let $A|kF$ be the factor algebra of A relative to the closed ideal kF. A/kF is known to be complete in the norm

$$\|f + kF\| = \inf \{\|g\|, g = f + h, h \in kF\}.$$

For an $x \in F$, we define on A/kF, the functional L_x by

$$L_x(f + kF) = f(x).$$

L_x is well defined, linear and multiplicative on A/kF. From Proposition 1.1 we get $\|L_x\| \leqslant 1$, hence $|f(x)| \leqslant \|f + kF\|$ for any $x \in F$ and so

$$\|f|K\| \leqslant \|f + kF\|. \tag{4.1.1.}$$

Let T be the restriction operator with values in $A|F$ defined on A by

$$Tf = f \mid F.$$

Clearly T is a linear, multiplicative map from A onto $A|F$ with Ker $T = kF$. Thus T induces an isomorphism $\widetilde{T}$ between A/kF and $A|F$. From (4.1.1) we get

$$\|\widetilde{T}(f + kF)\| = \|f|\,K\| \leqslant \|f + kF\| \tag{4.1.2}$$

and the closed graph theorem asserts that $\widetilde{T}$ is a topological isomorphism if and only if $A|F$ is closed.

Concerning the restriction algebra $A|F$ there arise the following questions: when is $A|F$ closed, or equivalently, under which conditions is the isomorphism between A/kF and $A|F$ topological? When is this isomorphism an isometry? When does $A|F = C(F)$ hold?

The conditions we give below will be expressed in duality terms. Even if some of the theorems are simple reformulations, they help to establish quite concrete results.

We recall now some notations. $\mathcal{M}(X)$ is the dual space of $C(X)$ and is considered as a space of measures. $\mathcal{M}(F)$ is naturally embedded in $\mathcal{M}(X)$ and for any $\mu \in \mathcal{M}(X)$ we can define the restriction measure $\mu_F = \chi_F \mu$.

We denote by $A^{\perp} = \{\mu \in \mathcal{M}(X)\colon\ \mu(f) = 0,\ f \in A\}$ and similarly, $(A|F)^{\perp}$. The dual of A (respectively $A|F$) can be identified with $\mathcal{M}(X)/A^{\perp}$ (respectively $\mathcal{M}(F)/(A|F)^{\perp}$) and the norms on these last spaces are those of a quotient space. Let T^* be the adjoint of T. We have

$$T^*\,(\nu + (A|F)^{\perp}) = (\nu + A^{\perp}) \qquad (\nu \in \mathcal{M}(F)).$$

Since for any $\nu \in \mathcal{M}(F) \bigcap A^{\perp}$ we have $\nu \in (A|F)^{\perp}$, T^* is an injection and

$$\|\,\nu + A^{\perp}\| \leqslant \|\,\nu + (A|F)^{\perp}\| \qquad (\nu \in \mathcal{M}(F)).$$

From the closed graph theorem we obtain T^* has a closed graph if and only if T^{*-1} is continuous. Since T has a closed graph if and only if T^* has a closed graph, we have the following

Theorem 4.1. *The algebra $A|F$ is closed in $C(F)$ if and only if there exists a $c \geqslant 1$ such that*

$$\|\nu + A|F)^{\perp} \| \leqslant c \| \nu + A^{\perp}\| \qquad (\nu \in \mathcal{M}(F)). \tag{4.1.3}$$

$c = 1$ if and only if $\widetilde{T}$ is an isometrical isomorphism between $A|F$ and $A|kF$.

The proof follows from the above considerations. ◇

Theorem 4.2. *The algebra $A|F$ is closed if and only if there is a $k < 1$ such that*

$$\|\mu_F + (A|F)^{\perp}\| \leqslant k\|\mu\| \qquad (\mu \in A^{\perp}), \tag{4.1.4}$$

$A|F$ is isometrically isomorphic to A/kF if and only if $k = 1/2$.

Proof. If $A|F$ is closed, then from (4.1.3) we obtain

$$\|\mu_F + (A|F)^{\perp}\| \leqslant c\|\mu\| - c\|\mu_F\|$$

for any $\mu \in A^{\perp}$.

Hence

$$(1 + c)\, \|\mu_F + (A|F)^{\perp}\| \leqslant \|\mu_F + (A|F)^{\perp}\| + c\|\mu_F\| \leqslant c\|\mu\|.$$

Writing $k = c/(1 + c)$ we have $k < 1$ and

$$\|\mu_F + (A|F)^{\perp}\| \leqslant k\|\mu\| \qquad (\mu \in A^{\perp}).$$

Obviously $k = 1/2$ if and only if $c = 1$.

We assume now (4.1.4) to be true. We prove first that there exists a $c \geqslant 1$ for which

$$\|\mu_F + (A|F)^{\perp}\| \leqslant c\|\mu_{X-F}\| \qquad (\mu \in A^{\perp}). \tag{4.1.5}$$

Let c_0 be the smallest real number ($c_0 = \infty$ if there is no such number) which satisfies (4.1.5) and $r < c_0$. There exists $\mu \in A^{\perp}$ such that

$$r\|\mu_{X-F}\| < \|\mu_F + (A|F)^{\perp}\|.$$

For any $t > 0$ we choose $\nu \in (A|F)^{\perp}$ with

$$\|\mu_F + \nu\| < \|\mu_F + (A|F)^{\perp}\| + t.$$

Let $\tilde{\mu} = \mu + \nu$. Then $\tilde{\mu} \in A^{\perp}$ and

$$\|\tilde{\mu}\| = \|\mu_F + \nu\| + \|\mu_{X-F}\| < \|\mu_F + (A|F)^{\perp}\| + t + \frac{1}{r}\|\mu_F + (A|F)^{\perp}\| \leqslant$$

$$\leqslant k\|\mu\| + \frac{k}{r}\|\mu\| + t = k\,\frac{1+r}{r}\|\mu\| + t.$$

There follows

$$k > (\|\tilde{\mu}\| - t)\frac{r}{\|\mu\|\,(1+r)} \geqslant 1 - \frac{t}{\|\mu\|}\,\frac{r}{1+r} \geqslant \frac{r}{r+1}.$$

If $c_0 = \infty$ then $k \geqslant 1$, which is impossible. Hence there is a $c < \infty$ for which (4.1.5) holds.

Let $r > c$, $\nu \in \mathcal{M}(F)$ and $\mu \in A^{\perp}$. From (4.1.5) we get

$$\|\mu_F + (A|F)^{\perp}\| \leqslant C\|\mu_{X-F}\|$$

and therefore there is $\tilde{\mu} \in (A|F)^{\perp}$ such that

$$\|\mu_F + \tilde{\mu}\| \leqslant r\|\mu_{X-F}\|.$$

We have

$$r\,\|\nu + \mu\| = r\|\nu + \mu_F + \mu_{X-F}\| = r\|\nu + \mu_F\| + r\|\mu_{X-F}\| \geqslant$$

$$\geqslant \|\nu + \mu_F\| + \|\mu_F + \tilde{\mu}\| \geqslant \|\,\nu - \tilde{\mu}\| \geqslant \|\nu + (A|F)^{\perp}\|.$$

Since this inequality holds for any $r > c$, we have

$$\|\nu + (A|F)^{\perp}\| \leqslant c\|\,\nu + A^{\perp}\| \qquad (\nu \in \mathcal{M}(F))$$

i.e. (4.1.3). The theorem is proved. ◇

Corollary 4.3. *$A|F = C(F)$ if and only if there exists a $k < 1$ such that*

$$\|\mu_F\| \leqslant k\|\mu\| \qquad (\mu \in A^\perp).$$

$A|F = C(F)$ and both isometrically isomorphic to A/kF if and only if

$$\|\mu_F\| \leqslant \frac{1}{2}\|\mu\| \qquad (\mu \in A^\perp).$$

Corollary 4.4. *If $\mu \in A^\perp$ implies $\mu_F \in A^\perp$, then $A|F$ and A/kF are isometrically isomorphic.*

If for any $\mu \in A^\perp$ we have $\mu_F = 0$, then $A/F = C(F)$ and is isometrically isomorphic to A/kF.

4.2. The case of the intersections of peak sets

Proposition 4.5. *Let F be an intersection of peak sets for A. Then $A|F$ is isometrically isomorphic to A/kF.*

Proof. From (4.1.1) we know that

$$\|f|F\| \leqslant \|f + kF\| \qquad (f \in A).$$

Let U be the open set on which $|f| < \|f|F\| + \varepsilon$ for a fixed $f \in A$. Obviously, $F \subset U$ and therefore there is a peak set K such that $F \subset K \subset U$. Let $g \in A$ with $g = 1$ on K and $|g| < 1$ on $X - K$. Since for any n, $g^n f \in f + kF$, there results

$$\|f + kF\| \leqslant \limsup \|g^n f\| \leqslant \|f|F\| + \varepsilon. \diamond$$

Proposition 4.6. *Let F be an intersection of peak sets. For any function $g \in A|F$ there is an $f \in A$ such that $f = g$ on F and $\|f\| = \|g\|$.*

Proof. Let $g \in A|F$ with $\|g\| = 1$. Since $A|F$ and A/kF are isometrically isomorphic, there is an $h \in A$ such that $h = g$ on F and $\|h\| \leqslant 3/2$. Let $K = \{x \colon |h(x)| \leqslant 1\}$. It is obvious that K is of G_δ-type, say $K = \bigcap V_n$. As $F \subset K \subset V_n$ then, for any n there exists a peak set K_n such that $F \subset K_n \subset V_n$.

Write $K_0 = \bigcap K_n$; K_0 is a peak set. Let $p \in A$ with $\|p\| = 1$ and

$$K_0 = \{x \in X: p(x) = 1\}.$$

Let

$$F_n = \{x: 1 + 2^{-n-1} \leqslant |h(x)| \leqslant 1 + 2^{-n}\}.$$

$K_0 \subset K$ and hence $F_n \cap K_0 = \varnothing$ and therefore for any n we find an r_n such that $|p^{r_n}| < 2^{-2n}$ on F_n. We write

$$f = \Sigma\, 2^{-n} p^{r_n} h.$$

Obviously $f \in A$. For $x \in F$ we have

$$f(x) = \Sigma\, 2^{-n} h(x) = h(x) = g(x).$$

If $|h(x)| \leqslant 1$ then $|f(x)| \leqslant 1$. Let $x \in X$ with $|h(x)| > 1$. Then, there is an m such that $x \in F_m$ and we get

$$|f(x)| \leqslant |h(x)|(\sum_{n \neq m} 2^{-n} + 2^{-m}|p^{r_m}(x)|) \leqslant (1 + 2^{-m})(1 - 2^{-m} + 2^{-3m}) =$$

$$= 1 - 2^{-2m} + 2^{-3m} + 2^{-4m} = 1 - (2^{2m} - 2^m - 1)/2^{4m} < 1.$$

Hence $\|f\| \leqslant 1$ on X and so $\|f\| = \|g\| = 1$.
The proposition is proved. ◇

Corollary 4.7. *Let F be an intersection of peak sets and K a compact set such that $K \cap F = \varnothing$. For any $\varepsilon > 0$ and $g \in A|F$ there exists an $f \in A$ with $f = g$ on F and $|f| < \varepsilon$ on K.*

Proposition 4.8. *Let F be an intersection of peak sets. For any $g \in A$ and $\varepsilon > 0$ there is an $f \in A$ with $\mathrm{Re} f = \mathrm{Re} g$ on F and*

$$\sup_X |\mathrm{Re} f| \leqslant \sup_F |\mathrm{Re}\, g| + 2\varepsilon.$$

Proof. Let τ be the conformal mapping of the domain $\{z \in C: : |z| < 1\}$ onto the domain $\{z \in C: |z| < 1, \ -\varepsilon < \mathrm{Im}\ z < \varepsilon\}$ for which $\tau(0) = 0$, $\tau(1) = 1$. We choose $\delta > 0$ such that τ maps the circle $\{z \in C: |z| < \delta\}$ into the circle $\{z \in C: |z| < \varepsilon\}$.

Let U be a neighbourhood of F on which

$$|\text{Re}g| < \sup_F |\text{Re}g| + \varepsilon.$$

According to Corollary 4.7 there is an $h' \in A$ with $h' = 1$ on F and $|h'| < \delta$ on $X - U$. Let $h = \tau(h')$. Then $h \in A$, $|\text{Re}h| < \varepsilon$ on $X - U$ and $|\text{Re}h| \leqslant 1$ on X. Now, for $f = gh$ we have $\text{Re}f = \text{Re}g\ \text{Re}h - \text{Im}g\ \text{Im}h$ and therefore $\text{Re}f = \text{Re}g$ on F. Moreover

$$|\text{Re}f| \leqslant \sup_F |\text{Re}g| + 2\varepsilon$$

on U and $|\text{Re}f| < 2\varepsilon$ on $X - U$. ◇

Corollary 4.9. *If B is the closure of $\text{Re}A$ in $C_R(X)$ then $B|F$ is a closed subspace of $C_R(F)$. If $B_F = \{u \in B: u = 0$ on $F\}$ then $B|F$ and B/B_F are isometrically isomorphic.*

Proof. Let $u \in \text{Re}A$ and $v \in \text{Re}A$ with $u = v$ on F and $\|v\| < \sup_F |u| + \varepsilon$.

Since $v - u \in B_F$ we have

$$\|u + B_F\| \leqslant \|u + v - u\| = \|v\| \leqslant \sup_F |u| + \varepsilon$$

hence

$$\|u + B_F\| \leqslant \sup_F |u|.$$

On the other hand, there is a $v \in B_F$ such that

$$\|u + B_F\| \geqslant \|u + v\| - \varepsilon = \sup_X |u + v| - \varepsilon \geqslant \sup_F |u| - \varepsilon. \ ◇$$

Theorem 4.10. *Let F be an intersection of peak sets and $p \in C_R(X)$, $p > 0$ on X. If $g \in A|F$ is such that $|g| \leqslant p$ on F, then there exists an $f \in A$ with $f = g$ on F and $|g| \leqslant p$ on X.*

Proof. It is no loss of generality to assume that $\|p\| \leqslant 1$. Let $\alpha, \beta > 0$ be such that $\alpha < \inf_X p(x)$, $\beta \leqslant p(x) - \alpha$ for any $x \in X$. As follows from Proposition 4.6, there is an $f_1 \in A$ with $f_1 = g$ on F and $\|f_1\| = \|g\|$. Let $K_1 = \{x \in X: |f_1(x)| \geqslant p(x) + \beta/2^3\}$. We then find an $f_2 \in A$ with $f_2 = g$ on F, $\|f_2\| = \|g\|$ and $|f_2| < \alpha$ on K_1 (Corollary 4.7).

Thus we can construct by induction the sequence (f_n) of functions in A and the sequence (K_n) of compact sets in X such that $f_n = g$ on F, $\|f_n\| = \|g\|$, $|f_n| < \alpha$ on K_{n-1} and

$$K_{n-1} = \bigcup_{j=1}^{n-1} \{x \in X: |f_j(x)| \geqslant p(x) + \beta/2^{n+1}\}.$$

We write $f = \alpha/(2 - \alpha) \sum_{j=1}^{\infty} (\alpha^j/2^j)f_j$. Since $|f_n| \leqslant 1$ we get $f \in A$.

Let $x \in A$. If $x \in \bigcap_{n=1}^{\infty} K_n$, then $|f_n(x)| < \alpha$ for any n and therefore $|f(x)| < \alpha < p(x)$. If $x \notin K_{n+1} - K_n$ for any n, then $|f_j(x)| < p(x) + \frac{\beta}{2^{n+1}}$ for any n, hence $|f_j(x)| \leqslant p(x)$ for any j, and therefore $|f(x)| \leqslant p(x)$.

Suppose now that there exists an n such that $x \in K_{n+1} - K_n$. In this case $|f_j(x)| < p(x) + \beta/2^{n+1}$ for $j = 1, 2, \ldots, n$ and $|f_j(x) < \alpha$ for $j \geqslant n + 1$. Then we have

$$|f(x)| \leqslant \sum \frac{\alpha^j}{2^j} |f_j(x)| \leqslant (p(x) + \beta/2^{n+1}) \sum_{j=1}^{n} \frac{1}{2^j} + \alpha \sum_{j=n+1}^{\infty} \frac{1}{2^j} =$$

$$= (p(x) + \beta/2^{n+1}) (1 - 1/2^n) + (1/2^n) \alpha =$$

$$= p(x) + \beta(1/2^{n+1} - 1/2^{2n+1}) = \frac{p(x) - \alpha}{2^n} \leqslant$$

$$\leqslant p(x) + \beta(1/2^{n+1} - 1/2^{2n+1}) - \beta/2^n =$$

$$= p(x) + \beta(1/2^{n+1} - 1/2^{2n+1} - 1/2^n) \leqslant p(x),$$

which completes the proof. ◇

Note that the converse assertion is also true.

Proposition 4.11. *If* $A|F$ *is closed in* $C(F)$ *then* $(kF)^\perp = A^\perp + \mathcal{M}(F)$.

Proof. One has of course $A^\perp + \mathcal{M}(F) \subset (kF)^\perp$. If $f \in C(X)$ is orthogonal on $A^\perp + \mathcal{M}(F)$ then it belongs to A since it is orthogonal on $A^\perp$ and vanishes on F since it is orthogonal on $\mathcal{M}$(F), i.e. $f \in kF$. Therefore $A^\perp + \mathcal{M}(F)$ is weakly dense in $(kF)^\perp$. It is now sufficient to prove that $A^\perp + \mathcal{M}(F)$ is weakly closed in $\mathcal{M}(X)$. Hence we have to show that the unit sphere of $A^\perp + \mathcal{M}(F)$ is weakly closed in $\mathcal{M}(X)$.

Let $\mu \in \mathcal{M}(X)$ be in the weak closure of the unit sphere of $A^\perp + \mathcal{M}(F)$ in $\mathcal{M}(X)$. Obviously $\|\mu\| \leqslant 1$. Now let $(\mu_i + \nu_i)$ with $\mu_i \in A^\perp$, $\nu_i \in \mathcal{M}(F)$ be a generalized sequence in the unit sphere of $A^\perp + \mathcal{M}(F)$, weakly convergent to μ. Since $A|F$ is closed in $C(F)$, then according to Theorem 4.1 there exists $c \geqslant 1$ such that

$$\|\nu + (A|F)^\perp\| \leqslant c\|\nu + A^\perp\| \qquad (\nu \in \mathcal{M}(F)).$$

Hence, for any i there exists a measure $\lambda_i \in (A|F)^\perp$ with

$$\|\nu_i - \lambda_i\| \leqslant 2c\|\nu_i + \mu_i\| \leqslant 2c.$$

Therefore the generalized sequence $\{\nu_i - \lambda_i\}$ in $\mathcal{M}(F)$ has a limit point ν (a weak limit in $\mathcal{M}(X)$), which belongs to $\mathcal{M}(F)$. Then clearly the generalized sequence

$$\{\mu_i + \lambda_i\} = \{\mu_i + \nu_i\} - \{\nu_i - \lambda_i\}$$

has $\mu - \nu$ as limit point. Since μ_i and λ_i belong to $A^\perp$ which is weakly closed in $\mathcal{M}(X)$, $\mu - \nu \in A^\perp$, hence $\mu \in A^\perp + \mathcal{M}(F)$.

The proposition is proved. ◇

Theorem 4.12. *The following assertions are equivalent:*

(i) *F is an intersection of peak sets.*

(ii) *For any $\mu \in A^\perp$ we have $\mu_F \in A^\perp$.*

Proof. (i) → (ii). If there exists a $\mu \in A^\perp$ with $\mu_F \notin A^\perp$ then there is a $\mu \in A^\perp$ with $\|\mu_{X-F}\| = 1$ and $\|\mu_F + (A|F)^\perp\| \geqslant \varepsilon_0 > 0$.

Let $f \in A$ be such that $\|f\| \leqslant 1$ and

$$\int_F f d\mu = r > 2\varepsilon_0.$$

Let K be a compact set, disjoint of F, and such that $|\mu|(X - (F \cup K)) < \varepsilon_0$. According to Corollary 4.7 there is $g \in A$ with $\|g\| \leqslant 1$, $g = f$ on F and $|g| < \varepsilon_0$ on K. We have

$$0 = |\textstyle\int_X g d\mu| \geqslant |\int_F g d\mu| - |\int_K g d\mu| - |\int_{X-(F\cup K)} g d\mu| \geqslant$$

$$\geqslant r - \varepsilon_0 - \|g\| \int_{X-(F\cup K)} d|\mu| \geqslant r - 2\varepsilon_0 > 0$$

which is impossible.

(ii) → (i). Let $B = \{f \in A : f = \text{const. on } F\}$. $\hat{X}$ be the factor space obtained from X by reducing F to a point, and $\hat{B} = \{\hat{f} \in C(\hat{X})$: there is an $f \in A$ with $\hat{f}(\hat{x}) = f(x)\}$. Obviously $\hat{B}$ is a uniformly closed subalgebra of $C(\hat{X})$, which contains the constants. We prove now that $\hat{B}$ separates the points of $\hat{X}$.

Let $p \in \hat{X}$ be the point obtained from F and $\hat{x} \in \hat{X}$, $\hat{x} \neq p$. Suppose $\hat{f}(p) = \hat{f}(\hat{x})$ for any $f \in B$. This means that $f(x) = 0$ for any $f \in kF$, and hence $\varepsilon_x \in (kF)^\perp$.

From (ii) and Theorem 4.2 we get that $A|F$ is closed and from Proposition 4.11 we have

$$(kF)^\perp = A^\perp + \mathcal{M}(F) = (A^\perp)_{X-F} + \mathcal{M}(F)$$

where $(A^\perp)_{X-F} = \{\mu_{X-F} \text{ with } \mu \in A^\perp\}$.

Hence $\varepsilon_x \in (A^\perp)_{X-F} \subset A^\perp$. Then $f(x) = 0$ for any $f \in A$ which is impossible since A contains the constants.

Let now $\hat{x}_1 \neq p \neq \hat{x}_2$. If $\hat{f}(\hat{x}_1) = \hat{f}(\hat{x}_2)$ for any $\hat{f} \in \hat{B}$, then the measure $\varepsilon_{x_1} - \varepsilon_{x_2} \in (kF)^\perp$ and, since it is supported in $X - F$ we get $\varepsilon_{x_1} - \varepsilon_{x_2} \in (A^\perp)_{X-F} \subset A^\perp$. Therefore $f(x_1) = f(x_2)$ for any $f \in A$ which is impossible since A separates the points of X.

Then $\hat{B}$ is a function algebra on $\hat{X}$.

Let $\nu \in \hat{B}^\perp$. We define $\mu \in \mathcal{M}(X)$ by

$$\mu(\varphi) = \nu(\hat{\varphi})$$

for $\varphi \in C(X)$ which are constant on F.

$$\mu(f) = 0 \qquad (f \in A)$$

and then, extending to $C(X)$ by Hahn-Banach theorem.

Clearly $\mu \in A^{\perp}$ and (ii) gives $\mu_F \in A^{\perp}$. Then we have

$$\nu(\{p\}) = \nu(\chi_{(p)}) = \mu(\chi_F) = \mu_F(1) = 0.$$

Hence for any $\nu \in \hat{B}^{\perp}$ we have $\nu(\{p\}) = 0$.

Let $\nu \ll \varepsilon_p$. Then $\varepsilon_p - \nu \in \hat{B}^{\perp}$ therefore $(\varepsilon_p - \nu)(\{p\}) = 0$, that is $\nu = \varepsilon_p$.

From Theorem 2.8 p is an intersection of peak sets for $\hat{B}$ and then, obviously, F is an intersection of peak sets for A. ◇

Corollary 4.13. *(Rudin-Carleson Theorem). Let μ be a measure on X such that any $\nu \in A^{\perp}$ is absolutely continuous with respect to μ, and F a closed set of X with $\mu(F) = 0$.*

Let $p \in C(X)$ with $p > 0$ and $f \in C(F)$ with $|f| \leqslant p$ on F. There exists $g \in A$ such that $g = f$ on F and $|g(x)| \leqslant p(x)$ on X.

Proof. For any $\nu \in A^{\perp}$ one obviously has $\nu_F = 0$ and, according to Corollary 4.4 $A|F = C(F)$. From Theorem 4.12 F is an intersection of peak sets and therefore Theorem 4.10 completes the proof. ◇

Corollary 4.14. *(Bishop's Lemma). If F is an intersection of peak sets for A and $K \subset F$ is an intersection of peak sets for $A|F$, then K is an intersection of peak sets for A.* ◇

Proof. From Theorem 4.12 there results $(A^{\perp})_F \subset (A|F)^{\perp}$ and

$$[(A|F)^{\perp}]_K \subset [(A|F)|K]^{\perp} = [A|K]^{\perp}. \text{ Hence } (A^{\perp})_K = [(A^{\perp})]_K \subset$$

$$\subset [(A|F)^{\perp}]_K \subset [A|K]^{\perp}$$

and Theorem 4.12 completes the proof. ◇

4.3. Antisymmetry

A subset $E \subset X$ will be called a *set of antisymmetry* of A if any function in A, which is real on E is constant on E. The algebra A is said to be *antisymmetric* on X if X is a set of symmetry of A.

The closure of a set of antisymmetry is obviously a set of antisymmetry.

Proposition 4.15. *Any maximal set of antisymmetry is closed. Every $x \in X$ belongs to only one maximal set of antisymmetry.*

Proof. It is sufficient to notice that the union of two sets of antisymmetry with nonvoid intersection is a set of antisymmetry, and the singletons are sets of antisymmetry.

Therefore the family $\{E_i\}_{i \in \mathscr{I}}$ of maximal sets of antisymmetry is a family of disjoint closed sets which covers X:

$$X = \bigcup_{i \in \mathscr{I}} E_i.$$

Proposition 4.16. *Any maximal set of antisymmetry is an intersection of peak sets.*

Proof. Let E be a set of antisymmetry and K the intersection of all peak sets containing E. We show first that K is a set of antisymmetry.

Let $f \in A$ have real values on K. Then f is constant on E and therefore we may assume $\|f\| = 1$ and $f = 0$ on E; indeed, if any $f \in A$ with real values on K were constant on X, then X would be a set of antisymmetry and we would have nothing to prove.

$$\text{Let } N = \{x \in K : f(x) = 0\} \text{ and } g = 1 - f|K.$$

We have $g \in A|K$, $g = 1$ on N and $|g| < 1$ on $K - N$, N is then a peak set for $A|K$. According to Corollary 4.14, N is an intersection of peak sets for A. Since $E \subset N$, from the definition of K we get $K \subset N$, hence $f = 0$ on K.

Thus K is a set of antisymmetry and since $E \subset K$ is a maximal set of antisymmetry, $E = K$, i.e. E is an intersection of peak sets. ◇

Proposition 4.17. *(de Branges' Lemma). Let μ be a measure on X which is an extremal point of the unit ball of $A^{\perp}$. Then its support is a set of antisymmetry.*

Proof. Let K be the support of μ and let $f \in A$, real on K. We may assume $0 < f < 1$ on K. We write

$$N_1 = \int f \mathrm{d}|\mu|, \quad N_2 = \int (1 - f) \mathrm{d}|\mu|.$$

Then $N_1, N_2 > 0$ and

$$N_1 + N_2 = \int f \mathrm{d}\,|\mu| + \int (1 - f) \mathrm{d}\,|\mu| = \int \mathrm{d}\,|\mu| = \|\mu\| = 1.$$

Let $\mathrm{d}\mu_1 = N_1^{-1} f \mathrm{d}\mu$, $\mathrm{d}\mu_2 = N_2^{-1}(1 - f) d\mu$. We have $\|\mu_1\| = \|\mu_2\| = 1$, $\mu_1, \mu_2 \in A^{\perp}$ and $\mu = N_1\mu_1 + N_2\mu_2$.

Since μ is extremal we get $\mu = \mu_1 = \mu_2$, hence $f = N_1$ on K. ◇

Theorem 4.18. *(Shilov-Bishop); Let A be a function algebra on X. The space X admits a decomposition of the form $X = \bigcup_{i \in \mathcal{I}} E_i$, where $(E_i)_{i \in \mathcal{I}}$ is the family of maximal sets of antisymmetry of A. We have $E_i \cap E_j = \varnothing$ for $i \neq j$ and*

1) $A \mid E_i$ is closed and antisymmetric,

2) $f \in C(X)$ belongs to A if and only if for any $i \in \mathcal{I}$, $f \mid E_i \in A \mid E_i$.

Proof. We need only prove (2). Let $f \in C(X)$ with $f|E_i \in A|E_i$ for any $i \in \mathcal{I}$ and $\mu \in A^{\perp}$. It is sufficient to show that $\mu(f) = 0$. We may, of course, assume $\|\mu\| = 1$ and also, using the Krein-Milman Theorem (Proposition 1.4), we may assume that μ is an extremal point of the unit ball of $A^{\perp}$. According to de Branges' Lemma (Proposition 4.17), the support of μ is a set of antisymmetry, i.e. it is included in one of the E_i sets. Hence $\mu(f) = 0$.

Therefore the theorem is completely proved. ◇

If A is a function algebra on X, the decomposition $X = \bigcup_{i \in \mathcal{I}} E_i$ given by Theorem 4.18 will be called the *Shilov-Bishop decomposition of X* (with respect to A).

4.4 Some characterizations of *C(X)*

Let A be a function algebra on X. From the Shilov-Bishop decomposition theorem (Theorem 4.18) one can get some characterizations of $C(X)$ among the function algebras on X, according to the following principle:

Theorem 4.19. *Let (P) be a property which is defined on any function algebra on X, for any compact space X. It is assumed that: if a function algebra on X has the property (P) and A is antisymmetric,* *then X reduces to one point. Let A be a function algebra on X and $X = \bigcup_{i \in \mathscr{I}} E_i$ be the Shilov-Bishop decomposition of X with respect to A. If $A \mid E_i$ has property (P) for any $i \in \mathscr{I}$, then $A = C(X)$.*

Proof. According to the hypotheses, E_i reduces to a single point, for any i. Then it is clear that for any $f \in C(X)$ we have $f|E_i \in A \mid E_i$ and therefore Theorem 4.18 gives $f \in A$, that is $A = C(X)$. ◇

Theorem 4.20. *(Stone-Weierstrass). Let A be a function algebra on X. If A is symmetric, i.e. for any $f \in A$ we have $\bar{f} \in A$, then $A = C(X)$.*

Proof. Let A be a function algebra on X which is antisymmetric and symmetric on X. If X contained more than one point, then, since ReA separates X, there would exist a non-constant function $u \in \text{Re}A$, $u = \text{Re}f$ with $f \in A$. But A is symmetric, hence $u = \frac{1}{2}(f + \bar{f}) \in A$ and, on the other hand, since A is antisymmetric u would be constant on X. We have reached a contradiction and therefore X reduces to a single point. Now, since for any closed set $F \subset X$, $A \mid F$ is symmetric if A is symmetric, we have the conditions of Theorem 4.19, hence $A = C(X)$. ◇

Theorem 4.21. *Let A be a function algebra on X. The space ReA is uniformly closed in $C_R(X)$ if and only if $A = C(X)$.*

Proof. If $A = C(X)$, then obviously ReA is closed. The converse assertion will be proved by using Theorem 4.19.

Let A be an antisymmetric algebra on X such that ReA is closed. We fix an $x_0 \in X$ and put $(\text{Re}A)_{x_0} = \{u \in \text{Re}A : u(x_0) = 0\}$.

Since ReA is closed, $(\text{Re}A)_{x_0}$ is closed. For $u \in (\text{Re}A)_{x_0}$ there exists a unique $v \in (\text{Re}A)_{x_0}$, such that $u + iv \in A$. Indeed if $u = \text{Re}f$ with $f \in A$, then $v = \text{Im}f - \text{Im}f(x_0) \in (\text{Re}A)_{x_0}$ and $u + iv \in A$.

If $u + iv \in A$, $u + iv' \in A$ with v, $v' \in (\text{Re}A)$, then $v - v' = -i[u + iv) - (u + iv')] \in A$ and since A is antisymmetric, we get $v - v' = v(x_0) - v'(x_0) = 0$, i.e. $v = v'$.

We now define an operator T on $(\text{Re}A)_{x_0}$ with values in $(\text{Re}A)_{x_0}$ by $Tu = v$, with v constructed as above. T is a bounded operator. Indeed, using the closed graph theorem, it is sufficient to prove that if $u_n \in (\text{Re}A)_{x_0}$, $u_n \to u$ and $v_n = Tu_n$ converges to v, then $Tu = v$; this follows easily from the definition of T.

Assume now that X contains more than one point. Then there exists $g \in A$ with $g(x_0) = 0$ and $g \neq 0$. Let $g = u + iv$. From the antisymmetry we get $u \neq 0$, $v \neq 0$ and therefore there is an $x_1 \in X$, $x_1 \neq x_0$ such that $v(x_1) = \|v\|$. We put $z_1 = g(x_1)$. Let N be an arbitrary integer and Φ the conform mapping of the interior of the rectangle $R_{0'} = \{-\|u\| \leqslant x \leqslant \|u\|, \ -\|v\| \leqslant y \leqslant \|v\|\}$ on the interior of the rectangle $R_N = \{\|u\| \leqslant x \leqslant \|u\|, -N \leqslant y \leqslant N\}$ such that Φ can be extended to a homeomorphism between R and R_N and $\Phi(0) = 0$, $\Phi(z_1) = iN$. It is known that X may be obtained as a limit of polynomials on R. Therefore $h = \Phi(g) \in A$ and if $h = s + it$ then clearly $s, t \in (\mathrm{Re}A)_{x_0}' \cdot \|s\| \leqslant \|u\|$, $\|t\| = N$ and $Ts = t$. Since N is arbitrary, this contradicts the fact that T is bounded, hence X reduces to a single point.

Now let A be a function algebra on X such that ReA is closed and $X = \bigcup_{i \in \mathscr{I}} E_i$ is the Shilov-Bishop decomposition of X with respect to A. Since for any $i \in \mathscr{I}$, E_i is an intersection of peak sets, according to Corollary 4.9, Re$(A|E)$ is closed.

We thus have the conditions of Theorem 4.19 from which we conclude that $A = C(X)$. ◇

Theorem 4.22. *Let A be a function algebra on X. The space ReA is an algebra if and only if $A = C(X)$.*

Proof. If $A = C(X)$, then obviously Re$A = C_R(X)$ is an algebra.

Now let A be an antisymmetric algebra on X and assume ReA is an algebra. Let $x_0 \in X$; from antisymmetry we get that for any $u \in \mathrm{Re}A$ there exists a unique $f \in A$ such that Re$f = u$ and Im$f(x_0) = 0$. We define on ReA the norm $N(u) = \|f\|$, where $f \in A$, Re$f = u$ and Im$f(x_0) = 0$.

ReA becomes a real Banach space with respect to this norm. According to the closed graph theorem and the uniform boundedness principle we get

$$N(uu') \leqslant KN(u)N(u') \quad (u,u' \in \mathrm{Re}\, A).$$

We now show that for any $p \in \mathrm{Re}A$ with $p > 0$ we have $\log p \in \mathrm{Re}A$. Let $B = \{\varphi \in C(X)\colon \varphi = u + iv, \ v \in \mathrm{Re}A\}$. Since Re$A$ is an algebra, B is also an algebra. For $f \in B$ we put $N(f) = N(u) + N(v)$ and $\|f\|' = \sup_{0 \leqslant \theta \leqslant 2\pi} N(0^{i)} f)$.

One easily verifies that B is a Banach algebra with respect to a norm equivalent to $\|.\|'$ and A, $\bar{A} \subset B$, ReB = ReA. Let $\mathfrak{M}_A$, $\mathfrak{M}_B$ be the maximal ideal spaces of A and B respectively. If $\varphi \in \mathfrak{M}_B$

then $\varphi|A \in \mathfrak{M}_A$. Since $p \in \mathrm{Re}A$ and $p > 0$ on X, we get $p > 0$ on $\mathfrak{M}_A$. Now let $p = \frac{1}{2}(f + \bar{f})$ with $f \in A$. For any $\varphi \in \mathfrak{M}_B$ we have

$$p(\varphi) = \frac{1}{2}(f(\varphi) + \overline{f(\varphi)}) = \frac{1}{2}(f(\varphi|A) + \overline{f(\varphi|A)}) = p(\varphi|A) > 0.$$

Hence $p > 0$ on $\mathfrak{M}_B$. Then $\log p \in B$ and, since p is real, $\log p \in \mathrm{Re}B = \mathrm{Re}A$.

We now assume, that X does not reduce to a single point. Then there is a $g \in A$ with $g(x_0) = 0$ and $\|g\| = 1$. Let $x \in X$ be such that $\|g\| = g(x) = 1$. We choose an analytic Φ in $|z| < 1$, continuous on $|z| \leqslant 1$ such that $0 < \mathrm{Re}\,\Phi \leqslant 1$ in $|z| \leqslant 1$, $\mathrm{Im}\Phi(0) = 0$ and $\mathrm{Im}\,\Phi(g(x)) \geqslant n$, where n is a fixed but arbitrary natural number. Let $f = \Phi(g)$. Then $f \in A$ and $0 < \mathrm{Re}f \leqslant 1$ on X, $\mathrm{Im}f(x_0) = 0$ and $\|f\| \geqslant n$. From the above considerations we deduce that there exists $F \in A$ such that $\mathrm{Re}F = \log(\mathrm{Re}f)$. Let $V = \exp(^1/_2\, F)$; then $V \in A$ and $|V|^2 = \mathrm{Re}f$. Therefore $\|V\| \leqslant 1$.

By using the identity

$$(\mathrm{Re}z)^2 = \frac{1}{2}(\mathrm{Re}z^2 + |z|^2)$$

we get

$$(\mathrm{Re}V)^2 = \mathrm{Re}\left(\frac{1}{2}(V^2 + f)\right).$$

Now, since for any $h \in A$ we have

$$N(\mathrm{Re}h) \geqslant \|h\| - |\mathrm{Im}h(x_0)|$$

there results

$$N[(\mathrm{Re}V)^2] \geqslant \frac{1}{2}(\|V^2 + f\| - |\mathrm{Im}V^2(x_0)|) \geqslant \frac{1}{2}(n - 2)$$

as $\|f\| \geqslant n$ and $\|V^2\| \leqslant 1$.

On the other hand, $N((\mathrm{Re}\,V)^2) \leqslant K(N(\mathrm{Re}\,V))^2$ and $N(\mathrm{Re}\,V) \leqslant$ $\leqslant 2\|V\|^2 \leqslant 2$. Therefore

$$\frac{1}{2}(n-2) \leqslant N[(\mathrm{Re}\,V)^2] \leqslant 4K.$$

Since K is a fixed constant and n is arbitrary, we have a contradiction. Hence X reduces to a single point.

If A is a function algebra on X with $\mathrm{Re}A$ an algebra and F a closed subset of X, then it is clear that $\mathrm{Re}(A/F)$ is an algebra.

Therefore the conditions of Theorem 4.19 are satisfied and we have $A = C(X)$. ◇

Theorem 4.23 *Let A be a function algebra on X. If for any closed subset F of X, $A|F$ is closed, then $A = C(X)$.*

Proof. We show first that if F and K are two disjoint compact subsets of X then there exists $f \in A$ with $f|F = 1$ and $f|K = 0$. Let $x \in F$ and $y \in K$; since A separates X, there is a function $g \in A$ such that $g(x) = 1$ and $g(y) = 0$. Let V_x and W_x be two neighbourhoods of x and y respectively, such that $|g(V_x)| < 1/4$, $|1 - g(W_x)| < 1/4$.

Consider now a sequence p_n of complex variable polynomials which converges uniformly on $\{z: |z| \leqslant 1/4\} \cup \{z: |1-z| \leqslant 1/4\}$ to the characteristic function of the subset $\{z: |z| \leqslant 1/4\}$.

Then $p_n \circ g | \overline{V}_x \cup \overline{W}_x$ is a sequence of elements in $A|\overline{V}_x \cup \overline{W}_x$, which converges uniformly on $V_x \cup \overline{W}_x$ to 1 on V_x and 0 on W_x. Since $A|\overline{V}_x \cup \overline{W}_x$ is closed, there exists $e \in A$ with $e(V_x) = 1$, $e(W_x) = 0$.

We fix $y \in K$ and cover F with a finite number of neighbourhoods $V_1, \ldots, V_n$ constructed as above and let $W_1, \ldots, W_n$ and $e_1, \ldots, e_n$ be the corresponding neighbourhoods of y and functions in A. Then $e_1' = e_1 + e_2 - e_1 e_2$ has the value 1 on $V_1 \cup V_2$ and 0 on $W_1 \cap W_2$, $e_2' = e_1' + e_3 - e_1' e_3$ has the value 1 on $V_1 \cup V_2 \cup V_3$ and 0 on $W_1 \cap$ $\cap W_2 \cap W_3$, etc.

Thus we obtain a function f_y in A which takes the value 1 on F and 0 on a neighbourhood U_y of y. Let us cover K with a finite number of such neighbourhoods $U_1, \ldots, U_n$ and let $f_1, \ldots, f_n$ be the corresponding functions in A. Then $f = f_1, \ldots, f_n$ belongs to A and $f|F = 1$ and $f|K = 0$.

Now, let $x \in X$ with the following property (P): there is a neighbourhood U of x and a constant M such that for any closed subset $F \subset U$

and any function $f \in A$ which takes only the values 1 and 0 on F, we have $\|f + kF\| < M$.

We prove that for any closed $K \subset U$ we have $A \mid K = C(K)$. Let $\mu \in [A \mid K]^{\perp}$ be a measure on K, K_1 a closed subset of K, $\varepsilon > 0$ and K_0 a closed subset of $K - K_1$ such that $|\mu|[(K - K_1) - K_0] < \varepsilon/M$. Let $f \in A$ with $f|K_1 = 1$, $f|K_0 = 0$ and $\|f\| < M$. From the first part of the proof and from property (P) of x it follows that there exists such an f in A. Then

$$0 = \int f d\mu = \int_{K_1} d\mu + \int_{K_0} 0 \, d\mu + \int_{(K-K_1)-K_0} f d\mu.$$

Hence

$$|\mu(K_1)| = \Big|\int_{(K-K_1)-K_0} f d\mu\Big| < \varepsilon.$$

But $\varepsilon > 0$ is arbitrary and therefore $\mu(K_1) = 0$ for any $K_1 \subset K$, hence $\mu = 0$. There results $(A|K)^{\perp} = 0$, that is $A|K = C(K)$.

Assume now that A is antisymmetric on X and that there is an $x \in X$ with property (P). We choose $V_1 \subset \overline{V}_1 \subset V_2 \subset \overline{V}_2 \subset V_3 \subset \overline{V}_3 \subset \subset U$, neighbourhoods of x. From the above considerations we can find a function $f \in A$ such that $f(X - \overline{V}_3) = 0$ and $f(\overline{V}_2) = 1$, and a function $g \in A$ with $g|\overline{V}_3$ real, $g(\overline{V}_3 - V_2) = 0$ and $g(\overline{V}_1) = 1$.

Then it is clear that fg is a real non-constant function in A, which contradicts the antisymmetry of A. Therefore, no point of X has the property (P).

The next step is to show that X has a finite number of points. Assume that X contains an infinite sequence $\{x_n\}$ of distinct points. Since none of them has property (P), there exists a sequence U_n of disjoint neighbourhoods of x_n, a sequence of closed sets $F_n \subset V_n$ and a sequence f_n of functions which take only the values 1 and 0 on F_n and $\|f_n + kF_n\| \geq n$.

Let $F = \overline{\bigcup F_n}$. Since for $n \neq m$, $V_n \cap F_m = \varnothing$ there results $F = F_n \cup [\overline{\bigcup_{m \neq n} F_m}]$ and $F_n \cap [\bigcup_{m \neq n} F_m] = \varnothing$. Hence there exists a function $f \in A$ with $f(F_n) = 1$ and $f(F - F_n) = 0$. We then have

$$n \leq \|f_n + kF_n\| = \|ff_n + kF_n\| \leq \|ff_n + kF\|.$$

On the other hand, since $A|F$ is closed, the isomorphism between $A|F$ and $A|kF$ is a topological one, hence there is an M such that

$$\|f + kF\| \leqslant M \, \|f|F\| \qquad (f \in A).$$

But $\|ff_n| F\| = 1$ and we reach a contradiction; therefore X is finite.

A separates the points of X and is antisymmetric, hence X reduces to only one point.

Now if the function algebra A on X has the property enounced by the theorem, it is clear that for any closed set $F \subset X$ the algebra $A|F$ has the same property.

We have then under the conditions of Theorem 4.19 and $A = C(X)$. ◇

Notes

Theorems 4.1 and 4.2 together with their corollaries are to be found, under different forms, in E. Bishop [3], T. Gamelin [1] and I. Glicksberg [1]. The results of paragraph 4.2 leading to the abstract form (Theorem 4.10) of Rudin-Carleson interpolation theorem (Corollary 4.13) are given in the same works.

Different forms of the Shilov-Bishop decomposition theorem (Theorem 4.18) have been stated by G. E. Shilov [1] and E. Bishop [2]. In the present book, we follow the proof given by I. Glicksberg [1] using de Branges' Lemma (Proposition 4.17).

Theorems 4.19 and 4.21 belong to J. Wermer [3] and Theorem 4.22 to K. Hoffman and J. Wermer [1]. Theorem 4.23 is due to I. Glicksberg [2].

CHAPTER 5

H^p-Spaces

5.1. Definitions and basic lemmas

Let A be a function algebra on X and μ a positive measure on X. For $1 \leqslant p < \infty$ we denote by $H^p(\mathrm{d}\mu)$ the $L^p(\mathrm{d}\mu)$-closure of A, which is a subspace of $L^p(\mathrm{d}\mu)$. $H^\infty(\mathrm{d}\mu)$ will denote the weak closure of A in $L^\infty(\mathrm{d}\mu)$.

In this chapter we present a theory of $H^p(\mathrm{d}\mu)$-spaces in analogy with the classical theory of Hardy's H^p-classes.

We denote by $B(X)$ the set of Borel functions on X and by $M(X)$ the set of bounded Borel functions on X.

For a function $F \in L^1(\mathrm{d}\mu)$, $F \geqslant 0$, its geometrical mean will be defined as

$$J(F;\, \mathrm{d}\mu) = \exp\,[\int \log F\mathrm{d}\mu].$$

Obviously $0 \leqslant J(F;\, \mathrm{d}\mu) < \infty$ and $J(F;\, \mathrm{d}\mu) > 0$ if and only if $\log F \in L^1(\mathrm{d}\mu)$.

Lemma 5.1. *Let μ be a positive measure on X for which $\mu(X) = 1$ and $G \in L^1(\mathrm{d}\mu)$ is non-negative. Then the function*

$$\theta(t) = [\int F^t\mathrm{d}\mu]^{1/t} \qquad (0 < t \leqslant 1)$$

is monotonously increasing. If F is not constant, θ is a strictly increasing function. Moreover

$$J(F;\, \mathrm{d}\mu) = \lim_{t\to 0} \theta(t).$$

Proof. Let $0 < t \leqslant s \leqslant 1$. By applying Hölder inequality for F^t and 1 and for $p = s/t$, we find $\theta(t) \leqslant \theta(s)$. If $\theta(t) = \theta(s)$ for $t \neq s$, then Hölder inequality becomes an equality; therefore, F and 1 are linear dependent, i.e. F is constant.

The well-known inequality $e^x \geqslant 1 + x$ for any real x, gives $F^t = \exp [t \log F] \geqslant 1 + t \log F$ for any t, $0 < t \leqslant 1$, hence

$$\theta(t) = [\int F^t d\mu]^{1/t} \geqslant [1 + t \int \log F d\mu]^{1/t}, \qquad (0 < t \leqslant 1).$$

Therefore

$$\theta(t) \geqslant \exp [\int \log F \, d\mu] = J(F, d\mu), \qquad (0 < t \leqslant 1),$$

that is

$$J(F, d\mu) \leqslant \lim_{t \to 0} \theta(t).$$

Let us now prove the converse inequality. First, we may suppose that $F \geqslant \alpha$, $\alpha > 0$, since the general case can be obtained from this one by Beppo-Levi Theorem. We then take $F \geqslant 1$, hence $\log F \geqslant 0$. Let $\delta > 0$ and $0 < t \leqslant \delta \leqslant 1$. We have

$$F^t = 1 + t \log F + \sum_{k=2}^{\infty} \frac{t^k}{k!} (\log F)^k \leqslant 1 + t \log F + t\delta \sum_{k=0}^{\infty} \frac{1}{k!} (\log F)^k =$$

$$= 1 + t(\log F + \delta F).$$

Then

$$\int F^t d\mu \leqslant 1 + t(\int \log F \, d\mu + \delta \int F \, d\mu)$$

and, therefore,

$$\theta(t) \leqslant [1 + t(\int \log F \, d\mu + \delta \int F \, d\mu)]^{1/t}$$

and we have

$$\lim_{t \to 0} \theta(t) \leqslant J(F, d\mu) \exp [\delta \int F d\mu]$$

for any δ, $0 < \delta \leqslant 1$, that is

$$\lim_{t \to 0} \theta(t) \leqslant J(F, d\mu)$$

which completes the proof. ◇

Lemma 5.2. *Let T be a linear subspace of $M(X)$ such that $1 \in T$, L be a linear functional on T, with $L(1) = 1$, and μ be a positive measure on X such that $\mu(X) = 1$. If for p, $1 \leqslant p < \infty$ and any $F_1, F_2 \in T$ we have*

$$|L(F_1)L(F_2)|^p \leqslant \int |F_1F_2|^p d\mu,$$

then

$$L(F_1)L(F_2) = \int F_1F_2 d\mu.$$

Proof. Let $F \in T$ with $L(F) = 0$. For any complex number λ we have

$$1 = |L(1 - \lambda F)|^p \leqslant \int |1 - \lambda F|^p d\mu$$

$$1 = |L(1 - \lambda F)L(1 + \lambda F)|^p \leqslant \int |1 - \lambda^2 F^2|^p d\mu.$$

Therefore, for any $t > 0$ and any complex number λ we have

$$\int \frac{1}{t}[|1 + t\lambda F|^p - 1] d\mu \geqslant 0$$

$$\int \frac{1}{t}[|1 + t\lambda F^2|^p - 1] d\mu \geqslant 0.$$

On the other hand, for any t, $0 < t \leqslant 1$, and any complex number z one can easily verify that

$$\frac{1}{t}[|1 + tz|^p - 1] \leqslant p|z|(1 + p|z|)^{p-1}$$

$$\lim_{t \to 0} \frac{1}{t}[|1 + tz|^p - 1] = p\mathrm{Re}z.$$

There results

$$\int \mathrm{Re}(\lambda F) d\mu \geqslant 0$$

$$\int \mathrm{Re}(\lambda F^2) d\mu \geqslant 0.$$

for any complex number λ; this is possible only if

$$\int F d\mu = \int F^2 d\mu = 0.$$

Consider now an arbitrary $F \in T$. Then

$$\int F\mathrm{d}\mu = \int [F - L(F) + L(F)]\mathrm{d}\mu = \int [F - L(F)]\mathrm{d}\mu + L(F) = L(F)$$

$$\int F^2\mathrm{d}\mu = \int [F - L(F) + L(F)]^2\mathrm{d}\mu = \int [F - L(F)]^2\mathrm{d}\mu +$$
$$+ 2L(F) \int [F - L(F)]\mathrm{d}\mu + [L(F)]^2 = L(F)^2.$$

For $F_1, F_2, \in T$ we have

$$2 \int F_1F_2\mathrm{d}\mu = \int (F_1 + F_2)^2\mathrm{d}\mu - \int F_1^2\mathrm{d}\mu - \int F_2^2\mathrm{d}\mu =$$
$$= [L(F_1 + F_2)]^2 - [L(F_1)]^2 - [L(F_2)]^2 = 2L(F_1)L(F_2).$$

The lemma is thus proved. ◇

Let $\mathfrak{M}$ be the maximal ideal space of A. Until the end of this chapter we shall fix an element $\varphi \in \mathfrak{M}$, that is a non-zero multiplicative linear functional on A. Then it is known that there is a representing measure μ for φ, with support in X, i.e. a positive measure μ on X such that

$$f(\varphi) = \int f\mathrm{d}\mu \qquad (f \in A).$$

Let S be a linear subspace of $B(X)$ and σ a linear functional on S. The couple $[S, \sigma]$ will be called *admissible* (relative to A and φ) if for any $f \in A$ and $h \in S$ we have $fh \in S$ and

$$\sigma(fh) = \varphi(f)\,\sigma(h) \qquad (f \in A, h \in S).$$

For an admissible couple $[S, \sigma]$, a positive measure μ on X and $p \in [1, \infty)$, we denote by

$$D^p(S, \sigma, \mathrm{d}\mu) = \inf\{\int |h|^p \mathrm{d}\mu, h \in S, \sigma(h) = 1\}.$$

We have $D^p(S, \sigma, \mathrm{d}\mu) \geqslant 0$, $D^p(S, \sigma, \mathrm{d}\mu) < \infty$ for $S \subset L^p(\mathrm{d}\mu)$ and the function $p \to D^p(S, \sigma, \mathrm{d}\mu)$ is right continuous on $[1, \infty)$.

The couple $[A, \varphi]$ is obviously admissible. We write

$$D^p(\mathrm{d}\mu) = D^p(A, \varphi, \mathrm{d}\mu).$$

Proposition 5.3. *Let μ be a positive measure on X with $\mu(X) = 1$. Then μ is a representing measure for φ if and only if $D^p(\mathrm{d}\mu) = 1$.*

Proof. Let μ be a representing measure for φ. As $\varphi(1) = 1$, there results

$$D^p(\mathrm{d}\mu) \leqslant \int \mathrm{d}\mu = 1.$$

On the other hand, if $f \in A$ and $\varphi(f) = 1$, then

$$1 = \varphi(f) = \int f\mathrm{d}\mu \leqslant \int |f|\, \mathrm{d}\mu \leqslant [\int |f|^p]^{1/p},$$

hence $D^p(\mathrm{d}\mu) \geqslant 1$.

Assume now $D^p(\mathrm{d}\mu) = 1$. For any $f \in A$, with $\varphi(f) \neq 0$, we have

$$1 = D^p(\mathrm{d}\mu) \leqslant \int \left| \frac{f}{\varphi(f)} \right|^p \mathrm{d}\mu,$$

hence

$$|\varphi(f)| \leqslant \int |f|^p \mathrm{d}\mu \qquad (f \in A).$$

From Lemma 5.2 there results

$$\varphi(f) = \int f\mathrm{d}\mu, \qquad (f \in A)$$

i.e. μ is a representing measure for φ. ◇

Lemma 5.4. *Let $[S, \sigma]$ be an admissible couple with $S \subset M(X)$, let μ be a positive measure on X and let $F \in L^1(\mathrm{d}\mu)$ be non-negative. Let $1 < p,\ q < \infty$ with $1/p + 1/q = 1$, and assume $D^p(S, \sigma, \mathrm{d}\mu) > 0$. Then*

1) *There is only one function $Q \in L^q(\mathrm{d}\mu)$ such that*

$$\int |Q|^q \mathrm{d}\mu = 1, \quad \int hQ^{1/p}\mathrm{d}\mu = \sigma(h)\, [D^p(S, \sigma, F\mathrm{d}\mu)]^{1/p} \qquad (h \in S).$$

2) $\mathrm{d}\lambda = |Q|^q\mathrm{d}\mu$ *is a representing measure for* φ.
3) *The function $P \in L^p(\mathrm{d}\mu)$ defined by $PQ \geqslant 0$ and*

$$|P|^p = D^p(S, \sigma, F\mathrm{d}\mu)\, |Q|^q$$

belongs to the $L^p(\mathrm{d}\mu)$-closure of the set

$$\{hF^{1/p}\colon h \in S,\ \sigma(h) = 1\}.$$

Proof. We have $S \subset L^\infty(\mathrm{d}\mu)$. We define, on the subspace $\{hF^{1/p}\colon h \in S\}$ of $L^p(\mathrm{d}\mu)$, the functional L by

$$L(hF^{1/p}) = [D^p(S,\sigma, F\mathrm{d}\mu)]^{1/p}\sigma(h).$$

Then

$$L(hF^{1/p}) = [D^p(S, \sigma, F\mathrm{d}\mu)]^{1/p}\sigma(h) \leqslant [\int |h|^p F\mathrm{d}\mu]^{1/p} =$$

$$= \|hF^{1/p}\|_{L^p(d\mu)}.$$

Therefore, L can be extended to a linear functional on $L^p(\mathrm{d}\mu)$, of norm less than 1. There exists then a function $Q \in L^q(\mathrm{d}\mu)$ such that

$$\int hQF^{1/p}\,\mathrm{d}\mu = [D^p(S, \sigma, F\mathrm{d}\mu)]^{1/p}\sigma(h) \qquad (h \in S) \tag{5.1.1}$$

and

$$\int |Q|^q \leqslant 1. \tag{5.1.2}$$

From (5.1.1), for any $f \in A$ with $\varphi(f) = 1$ and $h \in S$ with $\sigma(h) = 1$, we obtain

$$[D^p(S, \sigma, F\mathrm{d}\mu)]^{1/p} = \int fhQF^{1/p}\mathrm{d}\mu \leqslant$$

$$\leqslant [\int |f|^q |Q|^q\mathrm{d}\mu]^{1/q}[\int |h|^p F\mathrm{d}\mu]^{1/p}$$

and therefore

$$[\int |f|^q |Q|^q\mathrm{d}\mu]^{1/p} \geqslant \left[\frac{D^p(S, \sigma, F\mathrm{d}\mu)}{\int |h|^p F\mathrm{d}\mu}\right]^{1/p}$$

for any $f \in A$ with $\varphi(f) = 1$ and $h \in S$ with $\sigma(h) = 1$.

There results

$$\int |f|^q |Q|^q\mathrm{d}\mu \geqslant 1. \tag{5.1.3}$$

In particular we have

$$\int |Q|^q\mathrm{d}\mu \geqslant 1$$

which, combined with (5.1.3), yields

$$\int |Q|^q\mathrm{d}\mu = 1. \tag{5.1.4}$$

From (5.1.3) we get

$$D^q(|Q|^q\mathrm{d}\mu) = \inf\{\int |f|^q |Q|^q\mathrm{d}\mu;\ f \in A,\ \varphi(f) = 1\} \geqslant 1$$

and, since

$$D^q(|Q|^q\mathrm{d}\mu) \leqslant \int |Q|^q\mathrm{d}\mu = 1$$

there follows $D^q(|Q|^q \, d\mu) = 1$. Then, according to Proposition 5.3, $d\lambda = |Q|^q \, d\mu$ is a representing measure for φ.

Let Ω be the set of all functions of $L^q(d\mu)$ with the following property: considered as functionals on $L^p(d\mu)$ they extend L and their norm is at most equal to 1.

From (5.1.4) Ω results a closed set in $\{Q \in L^q: \|Q\| = 1\}$. As Ω is a convex set and the points of $\{Q \in L^q: \|Q\| = 1\}$ are extremal points for $\{Q \in L^q: \|Q\| \leqslant 1\}$, Ω reduces to only one point. This proves the uniqueness of the function Q.

1) and 2) are completely proved. We now prove 3).

Let T be the subspace of functions $g \in L^q(d\mu)$ which satisfy

$$\int ghF^{1/p} \, d\mu = 0 \qquad (h \in S, \sigma(h) = 0).$$

One can then easily verify that

$$\sigma(h_1) \int gh_2 F^{1/p} \, d\mu = \sigma(h_2) \int gh_1 F^{1/p} \, d\mu \qquad (g \in T, h_1, h_2 \in S)$$

and if we put

$$L(g) = \int ghF^{1/p} \, d\mu \qquad (h \in S, \sigma(h) = 1)$$

we obtain a linear functional L on T satisfying

$$L(g)\, \sigma(h) = \int ghF^{1/p} \, d\mu. \qquad (h \in S, g \in T)$$

We have

$$|L(g)| \leqslant \int |g| |h| \, F^{1/p} \, d\mu \leqslant \|g\|_{L^q(d\mu)} [\int |h|^p F d\mu]^{1/p}$$

for any $h \in S$ with $\sigma(h) = 1$, and, therefore,

$$|L(g)| \leqslant [D^p(S, \sigma, F d\mu)]^{1/p} \, \|g\|_{L^q(d\mu)}.$$

From (5.1.1) we get $Q \in T$ and

$$L(Q) = [D^p(S, \sigma, F d\mu)]^{1/p},$$

hence

$$\|L\| = [D^p(S, \sigma, F d\mu)]^{1/p}.$$

We now extend L to a functional on $L^q(d\mu)$, of the same norm, and let G be the function of $L^p(d\mu)$ representing it. Then

$$\int |G|^p d\mu = D^p(S, \sigma, F d\mu)$$

and

$$L(g) = \int gG\mathrm{d}\mu. \qquad (g \in T)$$

Since $Q \in T$ we have

$$\int QG\mathrm{d}\mu = L(Q) = [D^p(S, \sigma, F\mathrm{d}\mu)]^{1/p} = \|Q\|_{L^q(d\mu)}\|G\|_{L^q(d\mu)}.$$

Hence, Hölder inequality for Q and G is an equality, therefore $|G|^p = c|Q|^q$ where c is a constant and $QG \geqslant 0$. By integration we obtain $c = D^p(S, \sigma, F\mathrm{d}\mu)$; then $G = P$ where P is the function defined in 3).

Let now g be a function in $L^q(\mathrm{d}\mu)$, orthogonal to $\{hF^{1/p} : h \in S\} \subset \subset L^p(\mathrm{d}\mu)$. There results $g \in T$ and, therefore,

$$\int Pg\mathrm{d}\mu = L(g) = \int ghF^{1/p}\,\mathrm{d}\mu = 0. \qquad (h \in S, \sigma(h) = 1).$$

Thus, P belongs to the closure of the set $\{hF^{1/p} : h \in S\}$ in $L^p(\mathrm{d}\mu)$. Let $h_n \in S$ be such that the sequence $h_nF^{1/p}$ tends to P in $L^p(\mathrm{d}\mu)$. From (5.1.1) we get

$$\int h_nQF^{1/p}\,\mathrm{d}\mu = \sigma(h_n)\,[D^p(S, \sigma, F\mathrm{d}\mu)]^{1/p},$$

hence $\sigma(h_n)$ is convergent and

$$\lim_{n\to\infty} \sigma(h_n) = \frac{\int PQ\mathrm{d}\mu}{D^p(S, \sigma, F\mathrm{d}\mu)} = 1$$

The sequence $h'_n = \dfrac{h_n}{\sigma(h_n)} F^{1/p}$ then converges to P in $L^p(\mathrm{d}\mu)$ and, since $\sigma(h_n/\sigma(h_n)) = 1$, P belongs to the closure of $\{hF^{1/p} : h \in S, \sigma(h) = 1\}$ in $L^p(\mathrm{d}\mu)$.

The proof is thus complete. ◇

5.2. The theorem of F. and M. Riesz and Szegö theorem

Let X be the unit circle $\{z: |z| = 1\}$ of the complex plane and A the algebra of continuous functions on X which may be analytically extended in the interior (the standard algebra). Let m be the normalized Lebesgue measure on X, μ an arbitrary measure on X and $\mathrm{d}\mu =$

$= d\mu_a + d\mu_s$ the Lebesgue decomposition of μ with respect to m. The principal part of the well-known theorem of F. and M. Riesz is the following:

If μ is analytic, that is

$$\int z^n d\mu = 0 \qquad (n = 1,2,\ldots)$$

then $\mu_s \in A^\perp$.

Indeed, the measure $d\lambda = \bar{z} d\mu_s$ results analytic and, since $d\lambda$ is obviously singular with respect to m, we get

$$\int z^n d\mu_s = 0 \qquad (n = 0, 1, 2,\ldots).$$

Repeating this argument we obtain by induction

$$\int z^n d\mu_s = 0, \qquad (n = 0, \pm 1, \pm 2,\ldots)$$

and, therefore, $\mu_s = 0$.

This principal part of the F. and M. Riesz theorem can also be proved in the case of an arbitrary function algebra A on an arbitrary compact space X, using some special representing measures instead of Lebesgue measure.

Let us consider a function algebra A on X and fix an element $\varphi \in \mathfrak{M}$.

Proposition 5.5. *Let m be a representing measure for φ. The following assertions are equivalent:*

(1) *Any representing measure for φ is absolutely continuous with respect to m.*

(2) *Let μ be a positive measure on X and $d\mu = Fdm + d\mu_s$, $F \in L^1(dm)$, the Lebesgue decomposition of μ with respect to m. Then $D^p(d\mu) = D^p(Fdm)$.*

(3) *Let μ be a positive measure on X and $d\mu = Fdm + d\mu_s$, $F \in L^1(dm)$ the Lebesgue decomposition of μ with respect to m. Let $[S, \sigma]$ be an admissible couple, with $S \subset L^p(d\mu)$. Then $D^p(S, \sigma, d\mu) = D^p(S, \sigma, Fdm)$.*

Proof. (1) → (3). We obviously have

$$\int |h|^p F dm \leqslant \int |h|^p d\mu \qquad (h \in S),$$

and, therefore,

$$D^p(S, \sigma, Fdm) \leqslant D^p(S, \sigma, d\mu).$$

We observe that if $D^p(S, \sigma, d\mu) = 0$, then the converse inequality is obvious; hence, assume $D^p(S, \sigma, d\mu) > 0$. Since the function $p \to D^p(S, \sigma, d\mu)$ is right continuous, we may suppose $p > 1$.

We first assume $S \subset M(X)$. Let $p > 1$ and q with $1/p + 1/q = 1$. According to Lemma 5.4 there exists $Q \in L^q(d\mu)$ such that

$$\int |Q|^q d\mu = 1$$

and

$$\int hQ d\mu = [D^p(S, \sigma, d\mu)]^{1/p} \sigma(h) \qquad (h \in S).$$

Moreover, $d\lambda = |Q|^q \, d\mu$ is a representing measure for φ. From (1) we get λ absolutely continuous with respect to m, hence $Q = 0$ almost everywhere with respect to μ_s.

For any $h \in S$ with $\sigma(h) = 1$ we have

$$[D^p(S, \sigma, d\mu)]^{1/p} = \int hQ d\mu = \int hQF dm \leqslant$$
$$\leqslant [\int |h|^p F dm]^{1/p} [\int |Q|^q F dm]^{1/q} = [\int |h|^p F dm]^{1/p}.$$

Therefore $D^p(S, \sigma, d\mu) \leqslant D^p(S, \sigma, Fdm)$.

We now assume that $S \subset L^p(d\lambda)$ is arbitrary and let $V \in S$ be such that $\sigma(V) = 1$. We define

$$S^* = \{h \in M(X): hV \in S\}$$
$$\sigma^*(h) = \sigma(hV) \quad (h \in S).$$

We have $S^* \subset M(X)$ and we can easily verify that $[S^*, \sigma^*]$ is an admissible couple. Applying the first part of this proof to measure $|V|^p d\mu$ and couple $[S^*, \sigma^*]$ we obtain

$$D^p(S^*, \sigma^*, |V|^p d\mu) \leqslant D^p(S^*, \sigma^*, |V|^p F dm).$$

We have

$$D^p(S^*, \sigma^*, |V|^p d\mu) = \inf \{\int |hV|^p; h \in M(X), hV \in S, \sigma(hV) = 1\} \geqslant$$
$$\geqslant \inf \{\int |h|^p d\mu; h \in S, \sigma(h) = 1\} = D^p(S, \sigma, d\mu).$$

Hence

$$D^p(S, \sigma, d\mu) \leqslant D^p(S^*, \sigma^*, |V|^p d\mu) \leqslant D^p(S^*, \sigma^*, |V|^p F d\mu) \leqslant$$
$$\leqslant \int |V|^p F dm,$$

since $1 \in S^*$ and $\sigma^*(1) = 1$. Then

$$D^p(S, \sigma, d\mu) \leqslant D^p(S, \sigma, Fdm),$$

which completes the proof of implication (1) → (3).

(3) → (2) is obvious

(2) → (1). Let μ be a representing measure for φ and $d\mu = Fdm + d\mu_s$, $F \in L^1(dm)$, the Lebesgue decomposition of μ with respect to m. We have

$$1 = D^p(d\mu) = D^p(Fdm) \leqslant \int Fdm \leqslant \int Fdm + \mu_s(X) = \mu(X) = 1.$$

There results $\mu_s(X) = 0$ and, since μ_s is positive we get $\mu_s = 0$. The proposition is proved. ◇

Theorem 5.6. (F. and M. Riesz). *Let m be a representing measure for φ. The following assertions are equivalent:*

(1) *any representing measure for φ is absolutely continuous with respect to m.*

(2) *Let $\mu \in A^\perp$ and $d\mu = d\mu_a + d\mu_s$ be the Lebesgue decomposition of μ with respect to m. Then $\mu_s \in A^\perp$.*

(3) *Let μ be a measure on X such that*

$$\int f d\mu = 0 \qquad (f \in A, \int f dm = 0)$$

and $d\mu = d\mu_a + d\mu_s$ is the Lebesgue decomposition of μ with respect to m. Then $\mu_s \in A^\perp$.

Proof. (1) → (2). Let $\mu \in A^\perp$ and $d\mu = Fdm + d\mu_s$ be the Lebesgue decomposition of μ with respect to m. For any natural number n we put $d\mu_n = \left[1 + \frac{1}{n}\right] |F|\, dm + d|\mu_s|$. From Proposition 5.5 we get $D^2(d\mu_n) = D^2\left(\left[1 + \frac{1}{n} |F|\right] dm\right)$.

According to the definition of D^2 there exist $f_n \in A$ such that $\varphi(f_n) = 1$ and

$$\int |f_n|^2\, d\mu_n < D^2(d\mu_n) + \frac{1}{n}.$$

We have

$$\int |f_n|^2 \, dm + \frac{1}{n} \int |f|^2 |F| \, dm + \int |f_n|^2 \, d|\mu_s| \leqslant \tag{5.2.1}$$

$$\leqslant D^2(d\mu_n) + \frac{1}{n} = D^2\left(\left[1 + \frac{1}{n}|F|\right] dm\right) + \frac{1}{n} \leqslant$$

$$\leqslant 1 + \frac{1}{n} \int |F| \, dm + \frac{1}{n} = 1 + \frac{1}{n} K,$$

where

$$K = \frac{1}{n}(1 + \int |F| \, dm).$$

On the other hand

$$\int |f_n|^2 \, dm + \frac{1}{n} \int |f_n|^2 |F| \, dm \geqslant D^2([1 + |F|] \, dm)$$

and from (5.2.1) we obtain

$$\int |f_n|^2 \, d\mu_s < \frac{1}{n}. \tag{5.2.2}$$

According to (5.21) we also have

$$\int |f_n|^2 \, dm + \frac{1}{n} \int |f_n^2| \, |F| \, dm < 1 + \frac{1}{n} K$$

and, since

$$\int |f_n|^2 \, dm \geqslant D^2(dm) = 1$$

there results

$$\int |f_n|^2 |F| \, dm < K. \tag{5.2.3}$$

Once more from (5.2.1), we get

$$\int |f_n|^2 \, dm < 1 + K$$

and, using also (5.2.3) we deduce

$$\int |f_n|^2 (1 + |F|) \, dm < 1 + 2K. \tag{5.2.4}$$

Therefore, the sequence (f_n) is bounded in $L_p(\mathrm{d}m + |F|\,\mathrm{d}m)$. We may assume (f_n) weakly convergent (we eventually consider a subsequence). Moreover, we have

$$\int |1 - f_n|^2 \,\mathrm{d}m = \int |f_n|^2 \,\mathrm{d}m + 1 - 2\mathrm{Re} \int f_n \mathrm{d}m = \int |f_n|^2 \mathrm{d}m - 1 < \frac{1}{n} K.$$

We used the fact that m is a representing measure for φ, $\varphi(f_n) = 1$, and the inequality (5.2.1). Hence, (f_n) is weakly convergent to 1 in $L^2(\mathrm{d}m + |F|\,\mathrm{d}m)$ and, therefore, the same is true in $L^2(|F|\,\mathrm{d}m)$.

Since $\mu \in A^{\perp}$, we have

$$0 = \int f f_n \mathrm{d}\mu = \int f f_n F \,\mathrm{d}m + \int f f_n \mathrm{d}\mu_s$$

for any $f \in A$. Since $\int f f_n F \mathrm{d}m$ converges to $\int f F \mathrm{d}m$ and, according to (5.2.2), $\int f f_n \mathrm{d}\mu_s$ converges to 0, we find

$$\int f F \,\mathrm{d}m = 0 \qquad (f \in A),$$

that is $F\,\mathrm{d}m \in A^{\perp}$. Therefore, $\mu_s \in A^{\perp}$ and (1) → (2) is proved.

(2) → (3). Consider a measure on X such that

$$\int f \mathrm{d}\mu = 0 \qquad (f \in A, \int f \mathrm{d}m = 0).$$

The measure $\mu - \mu(1)m$ then belongs to $A^{\perp}$ and from (2) there results that its singular part relative to m, which is obviously μ_s, belongs to $A^{\perp}$.

(3) → (1) is immediate. Indeed, any representing measure μ for φ satisfies the condition required in (3) and therefore $\mu_s \in A^{\perp}$. Since μ and m are positive, μ_s is also positive, hence $\mu_s = 0$.

The theorem is proved. ◇

We now establish Szegö theorem in the context of function algebras.

In the classical case, Szegö theorem established the distance in $L^2(F\,\mathrm{d}m)$, where m is the Lebesgue measure, $F \in L^1(\mathrm{d}m)$, $F \geqslant 0$, from 1 to the subspace generated by A_0 — the set of all functions in the standard algebra A which vanish in the origin — by the formula

$$\inf_{f \in A_0} \int |1 - f|^2 \, F \,\mathrm{d}m = \exp\left[\int \log F \,\mathrm{d}m\right].$$

If φ is the functional on A given by the value of functions at 0, then it is clear that

$$\inf_{f \in A_0} \int |1 - f|^2 F \, dm = D^2(F \, dm)$$

and, therefore, Szegö's formula may be written as

$$D^2(F dm) = J(F, dm).$$

In this form it will also be established in the general case.

Let A be a function algebra on X and φ a linear multiplicative functional on A.

Proposition 5.7. *Let m be a representing measure for φ such that any representing measure for φ, absolutely continuous with respect to m, coincides with m.*

Let $G \in L^1(dm)$, $G \geqslant 0$ and $d\mu = G dm$. Then for any $F \in L^1(d\mu)$, $F \geqslant 0$, $FG \in L^1(dm)$, any p, $1 \leqslant p < \infty$ and any admissible couple $[S, \sigma]$ with $S \subset L^p(d\mu)$, $S \subset L^p(Fd\mu)$ such that at least one of the numbers $D^p(S, \sigma, d\mu)$, $D^p(S, \sigma, Fd\mu)$ is different from zero, we have

$$J(F, dm) = \frac{D^p(S, \sigma, Fd\mu)}{D^p(S, \sigma, d\mu)}.$$

Proof. Since the function $p \to D^p$ is right continuous we may suppose $p > 1$. Let q be such that $1/p + 1/q = 1$. We firstly assume $F \geqslant \varepsilon > 0$. We now prove the inequality

$$J(F, dm) \leqslant \frac{D^p(S, \sigma, Fd\mu)}{D^p(S, \sigma, d\mu)}.$$

If $D^p(S,\sigma,d\mu) = 0$ then, according to the hypothesis, $D^p(S,\sigma,Fd\mu) > 0$, and therefore, in this case, the inequality is obvious. We may thus assume that $D^p(S, \sigma, d\mu) > 0$. We also assume, for the beginning, that $S \subset M(X)$. Since $D^p(S, \sigma, d\mu) \leqslant D(S, \sigma, Fd\mu)$, there follows that $D^p(S, \sigma, F^t d\mu) > 0$ for any t, $0 < t \leqslant 1$. According to Lemma 5.4 there exist $Q_t \in L^q(d\mu)$ and $P_t \in L^p(d\mu)$ such that $P_t Q_t \geqslant 0$, $|P_t|^p = D^p(S, \sigma, F^t d\mu) |Q_t|^q$. P_t belongs to the closure of $\{hF^{t/p}, h \in S, \sigma(h) = 1\}$ in $L^p(d\mu)$ and

$$\int |Q_t|^q d\mu = 1.$$

Moreover,

$$\int hQ_t F^{t/p}\,\mathrm{d}\mu = [D^p(S, \sigma, F^t\mathrm{d}\mu)]^{t/p}\sigma(h), \qquad (h \in S)$$

and $\mathrm{d}\lambda = |Q_t|^q\mathrm{d}\mu = |Q_t|^q F\,\mathrm{d}m$ is a representing measure for φ with $\mathrm{d}\lambda = \mathrm{d}m$ as $\mathrm{d}\lambda$ is absolutely continuous with respect to m.

For any $0 < s \leqslant t \leqslant 1$ we have

$$[D^p(S, \sigma, F^s\mathrm{d}\mu)]^{1/p}\sigma(h) = \int hF^{t/p}\left(\frac{1}{F}\right)^{1/p} Q_s\mathrm{d}\mu \quad (h \in S). \tag{5.2.5}$$

We choose a sequence (h_n), $h_n \in S$, with $\sigma(h_n) = 1$ and $h_n F^{t/p}$ converging to P_t in $L^p(\mathrm{d}\mu)$. If we write (5.2.5) for every h_n, we obtain at limit

$$[D^p(S, \sigma, F^s\mathrm{d}\mu)]^{1/p} = \int P_t\left(\frac{1}{F}\right)^{1/p} Q_s\mathrm{d}\mu. \tag{5.2.6}$$

Applying Hölder's inequality we get

$$\int P_t\left(\frac{1}{F}\right)^{\frac{t-s}{p}} Q_s\mathrm{d}\mu \leqslant \left[\int |P_t|^p\left(\frac{1}{F}\right)^{t-s}\mathrm{d}\mu\right]^{1/p} \int |Q_s|^q\,\mathrm{d}\mu =$$

$$= [D^p(S, \sigma, F^t\,\mathrm{d}\mu)]^{1/p}\left[\int\left(\frac{1}{F}\right)^{t-s} |Q_t|^q\mathrm{d}\mu\right]^{1/p} =$$

$$= [D^p(S, \sigma, F^t\,\mathrm{d}\mu)]^{1/p}\left[\int\left(\frac{1}{F}\right)^{t-s}\mathrm{d}m\right]^{1/p}$$

where we used the relation

$$\int |Q_s|^q\,\mathrm{d}\mu = 1,$$

and the fact that $\mathrm{d}\lambda = |Q_t|^q\mathrm{d}\mu$ is equal to m, and $|P_t|^p = D^p(S, \sigma, F^t\mathrm{d}\mu)\,|Q_t|^q$.

Then, from (5.2.6) we obtain

$$\frac{D^p(S, \sigma, F^s\,\mathrm{d}\mu)}{D^p(S, \sigma, F^t\,\mathrm{d}\mu)} \leqslant \int\left(\frac{1}{F}\right)^{t-s}\mathrm{d}m. \tag{5.2.7}$$

Hence, for $0 < s < t \leqslant 1$ we have

$$\left[\frac{D^p(S, \sigma, F^s\mathrm{d}\mu)}{D^p(S, \sigma, F^t\mathrm{d}\mu)}\right]^{\frac{1}{t-s}} \leqslant \left[\int\left(\frac{1}{F}\right)^{t-s}\mathrm{d}m\right]^{\frac{1}{t-s}}.$$

According to Lemma 5.1.2 there results

$$\limsup_{t \to s} \left[\frac{D^p(S, \sigma, F^s d\mu)}{D^p(S, \sigma, F^t d\mu)}\right]^{\frac{1}{t-s}} \leqslant J\left(\frac{1}{F}, dm\right)$$

for any $0 \leqslant s \leqslant t \leqslant 1$. In this case it is known that

$$\left[\frac{D^p(S, \sigma, F^s d\mu)}{D^p(S, \sigma, F^t d\mu)}\right]^{\frac{1}{t-s}} \leqslant J\left(\frac{1}{F} dm\right) = \frac{1}{J(F, dm)},$$

for any $0 \leqslant s < t < 1$. If we write the last inequality for $t = 1$ and $s = 0$ there results

$$J(F, dm) \leqslant \frac{D^p(S, \sigma, F d\mu)}{D^p(S, \sigma, d\mu)}.$$

We now give up the restriction $S \subset M(X)$. For this, let $V \in S$ with $\sigma(V) = 1$. Since $S \subset L^p(d\mu)$ and $S \subset L^p(F d\mu)$, we have

$$\int |V|^p d\mu < \infty, \quad \int |V|^p F d\mu < \infty.$$

We write

$$S^* = \{h \in M(X): \ hV \in S\}$$

$$\sigma^*: \sigma^*(h) = \sigma(hV), \quad h \in S^*.$$

One easily verifies that $[S^*, \sigma^*]$ is an admissible couple, $1 \in S^*$ $\sigma^*(1) = 1$; we also have $S^* \subset M(X)$. Then

$$D^p(S^*, \sigma^*, |V|^p d\mu) =$$

$$= \inf \{\int |Vh|^p dm: \ h \in M(X), \ hV \in S, \ \sigma(hV) = 1\} \geqslant$$

$$\geqslant \inf \{\int |h|^p d\mu, \ h \in S, \ \sigma(h) = 1\} = D^p(S, \sigma, d\mu) > 0.$$

Since $F \in I^1(|V|^p d\mu)$, we get

$$D^p(S^*, \sigma^*, F|V|^p d\mu) = \inf \{|h|^p |F| |V|^p d\mu, \ h \in S^*, \ \sigma^*(h) = 1\} \leqslant$$

$$\leqslant \int |V|^p F d\mu \text{ since } 1 \in S^* \text{ and } \sigma^*(1) = 1.$$

By applying the first part of the proof to the couple $[S^*, \sigma^*]$, the measure $|V|^p d\mu = |V|^p G\, dm$ and the function F, we obtain

$$J(F, dm) \leqslant \frac{D^p(S^*, \sigma^*, F|V|^p d\mu)}{D^p(S^*, \sigma^*, |V|^p d\mu)} \leqslant \frac{\int |V|^p F d\mu}{D^p(S, \sigma, d\mu)}.$$

Since this inequality holds for any $V \in S$ with $\sigma(V) = 1$, there results

$$J(F, dm) \leqslant \frac{D^p(S, \sigma, Fd\mu)}{D^p(S, \sigma, d\mu)}. \tag{5.2.8}$$

We can easily see that $\frac{1}{F}$ and FG have the same properties as F and G, enounced in the theorem. Therefore, if we apply (5.2.8) to them, we obtain

$$\frac{1}{J(F, dm)} = J\left(\frac{1}{F}, dm\right) \leqslant \frac{D^p(S, \sigma, d\mu)}{D^p(S, \sigma, Fd\mu)},$$

that is

$$J(F, dm) \geqslant \frac{D^p(S, \sigma, Fd\mu)}{D^p(S, \sigma, d\mu)} \tag{5.2.9}$$

which, together with (5.2.8), gives

$$J(F, dm) = \frac{D^p(S, \sigma, Fd\mu)}{D^p(S, \sigma, d\mu)}. \tag{5.2.10}$$

By applying (5.2.10) to the function $F + \varepsilon$, and then tending to limit according to the theorem of monotone convergence, we get rid of the restriction $F \geqslant \varepsilon > 0$.

The proposition is proved. ◇

Theorem 5.8 (Szegö). *Let m be a representing measure for φ. The following assertions are equivalent.*

(1) *Any representing measure for φ, absolutely continuous with respect to m, is equal to m.*

(2) *For any $F \in L^1(dm)$, $F \geqslant 0$, we have*

$$D^p(F\, dm) = J(F, dm).$$

Proof. (1) → (2) obviously results from the preceding proposition.

Let us prove the converse implication. Let $d\mu = F dm$ with $F \in L^1(dm)$, $F \geqslant 0$, be a representing measure for φ, absolutely continuous with respect to m. We then have

$$1 = D^p(F\,dm) = J(F, dm) \leqslant \int F\,dm = \mu(X) = 1$$

where we used Proposition 5.3. and Lemma 5.1. There results

$$J(F, dm) = \int F\,dm = 1$$

and, according to Lemma 5.1, we get $F = 1$, that is $\mu = m$.

Theorem 5.9 (Szegö-Kolmogorov-Krein). *Let m be a representing measure for φ. The following assertions are equivalent.*

(1) *m is the only representing measure for φ.*

(2) *Let μ be a positive measure on X and $d\mu = F\,dm + d\mu_s$ its Lebesgue decomposition with respect to m. Then*

$$D^p(d\mu) = J(F, dm).$$

Proof. The proof follows as an immediate consequence of Theorem 5.8 and Proposition 5.5. ◇

5.3. The factorization theorem

Let A be a function algebra on X and φ a non-zero linear and multiplicative functional on A. Throughout this chapter we shall denote by m a representing measure for φ such that any representing measure for φ, absolutely continuous with respect to m, is identical with m. As we have seen, this is equivalent to the fact that Szegö theorem holds for m.

We denote by $K(dm)$ the set of functions $h \in L^1(dm)$ with the property

$$\int fh\,dm = \varphi(f) \int h\,dm \qquad (f \in A).$$

Proposition 5.10. *Jensen's inequality*

$$\left| \int h dm \right| \leqslant J(|h|, dm)$$

holds for any $h \in K(dm)$.

Proof. We have

$$J(|h|, dm) = D^1(|h|\,dm) =$$

$$= \inf\{\textstyle\int |f|\,|h|dm,\ f \in A,\ \varphi(f) = 1\} \geqslant \inf\{|\textstyle\int fh\,dm|,\ f \in A, \varphi(f) = 1\} =$$

$$= \inf\{|\varphi(f)\textstyle\int h\,dm|,\ f \in A,\ \varphi(f) = 1\} = |\textstyle\int hdm|.$$

A function $E \in K(dm)$ will be called an outer function if

$$\int Edm = J(|E|; dm) > 0.$$

Proposition 5.11. *Let* $F \in L^1(dm)$ *with* $F \geqslant 0$ *and* $J(F, dm) > 0$. *Then there exists an outer function* $E \in K(dm)$ *such that*

(1) $|E| = F$

(2) *Any function* $G \in K(dm)$ *which satisfies*

$$|G| \leqslant F,\ \int Gdm = J(F, dm)$$

is equal to E.

(3) *Let* $h \in L^1(dm)$ *with* $hE \in L^1(dm)$. *Then* $h \in K(dm)$ *if and only if* $hE \in K(dm)$. *In this case we have*

$$\int hEdm = \int hdm \int Edm.$$

Proof. Let S be the subspace of $L^1(dm)$ defined by

$$S = \{h \in L^1(dm)\colon h = fF,\ f \in A\}.$$

We define on S a functional L by

$$L(fF) = D^1(F\,dm)\,\varphi(f).$$

Of course L is linear on S. We have

$$|L(fF)| = D^1(Fdm)\,|\varphi(f)| \leqslant \int |f|Fdm = \|fF\|_{L^1(dm)}.$$

Therefore L may be extended to a linear functional of norm at most 1, on $L^1(dm)$. Thus, there exists $Q \in L^\infty(dm)$, with $|Q| \leqslant 1$, and

$$D^1(Fdm)\,\varphi(f) = \int fFQdm \qquad (f \in A).$$

Let $E = QF$. Then $E \in L^1(dm)$ and

$$\int Efdm = \int QFfdm = D^1(Fdm)\,\varphi(f) = \varphi(f)\,Edm,$$

that is $E \in K(\mathrm{d}m)$. Moreover,

$$\int E \mathrm{d}m = \int QF\, \mathrm{d}m = D^1(F\mathrm{d}m) = J(F, \mathrm{d}m) > 0$$

and, since obviously $|E| \leqslant F$, we obtain

$$J(F, \mathrm{d}m) = \int E \mathrm{d}m \leqslant J(|E|, \mathrm{d}m) \leqslant J(F, \mathrm{d}m).$$

Hence

$$\int E \mathrm{d}m = J(|E|, \mathrm{d}m) > 0$$

i.e. E is an outer function.

We now write

$$S = \{ h \in B(X) \colon hF \in K(\mathrm{d}m)\}$$

$$\sigma \colon \sigma(h) = \int hF \mathrm{d}m \qquad (h \in S).$$

S is a linear subspace of $B(X)$ and σ a linear functional on S. If $G \in K(\mathrm{d}m)$ with $|G| \leqslant F$ and

$$\int G \mathrm{d}m = J(F, \mathrm{d}m) > 0$$

then $G = hF$, $h \in S$, $|h| \leqslant 1$, since $J(F\, \mathrm{d}m) > 0$ yields $F \neq 0$ almost everywhere with respect to m. We then have

$$\sigma(h) = \int G \mathrm{d}m = J(F, \mathrm{d}m) > 0$$

hence $\sigma \not\equiv 0$.

We easily verify that $[S, \sigma]$ is an admissible couple. We have

$$D^1(S, \sigma, F\mathrm{d}m) = \inf \{ \int |h|\, F\mathrm{d}m, hF \in K(\mathrm{d}m), \int hF\mathrm{d}m = 1\} =$$
$$= \inf \{ \int |g|\, \mathrm{d}m, g \in K(\mathrm{d}m), \int g\mathrm{d}m = 1\} = 1.$$

From Proposition 5.7 we get

$$\frac{1}{D^1(S, \sigma, \mathrm{d}m)} = \frac{D^1(S, \sigma, F\mathrm{d}m)}{D^1(S, \sigma, \mathrm{d}m)} = J(F, \mathrm{d}m).$$

Hence we have $0 < D^1(S, \sigma, \mathrm{d}m) < \infty$ and

$$1 \leqslant J(F, \mathrm{d}m)\, D^1(S, \sigma, \mathrm{d}m),$$

which yields

$$|\sigma(h)| \leqslant J(F, \mathrm{d}m) \int |h|\, \mathrm{d}m \qquad (h \in S)$$

and therefore

$$\int hF\,dm \leqslant J(F, dm)\int |h|\,dm \qquad (h \in L^1(dm),\ hF \in K(dm)).$$

Thus there exists $Q \in L^\infty(dm)$ with $|Q| \leqslant 1$ and

$$\int h\,Fdm = J(F, dm)\int h\,Qdm, (h \in L^1(dm),\ hf \in K(dm)).$$

Let now $G \in K(dm)$ with $|G| \leqslant F$ and

$$\int Gdm = J(F, dm) > 0.$$

Consider $h \in B(X)$ with $G = hF$. Then $|h| \leqslant 1$ and

$$\int h\,Qdm = [J(F, dm)]\int h\,Fdm = 1.$$

There results $hQ = 1$ and $|h| = |Q| = 1$. Therefore, there is only one function $G \in K(dm)$ such that $|G| \leqslant F$.

$$\int Gdm = J(F, dm),$$

and which, moreover, verifies $|G| = F$. This function is obviously E. (1) and (2) are proved. We now prove point (3).

We first observe that

$$\int hFdm = \int Edm\int h\frac{F}{E}\,dm\ (h \in L(dm);\ hF \in K(dm)).$$

Let $h \in L^1(dm)$ such that $hE \in K(dm)$. Then

$$\int h\,Edm = \int \frac{hE}{F}\,F\,dm = \int Edm\int hdm.$$

If $f \in A$ with $\varphi(f) = 0$, then $fhE \in K(dm)$ and

$$\int fhEdm = 0.$$

Hence

$$0 = \int fhEdm = \int fhdm\int Edm,$$

that is

$$\int fhdm = 0,$$

and therefore $h \in K(dm)$.

Let now $h \in K(\mathrm{d}m)$ such that $hE \in L^1(\mathrm{d}m)$. We show at first that

$$\int hE\mathrm{d}m = \int h\,\mathrm{d}m \int E\mathrm{d}m.$$

In order to do this we consider

$$S = \{h \in B(X) : h \in K(\mathrm{d}m)\}$$
$$\sigma : \sigma(h) = \int h\mathrm{d}m \qquad (h \in S).$$

One easily verifies that $[S, \sigma]$ is an admissible couple, $S \subset L^1(\mathrm{d}m)$ and $D^1(S, \sigma, \mathrm{d}m) = 1$. Then, from Proposition 5.5 there results

$$J(F, \mathrm{d}m)\,|\int h\mathrm{d}m| \leqslant \int |h|\, F\,\mathrm{d}m \qquad (h \in K(\mathrm{d}m)).$$

Hence, there exists $Q \in L^\infty(\mathrm{d}m)$, $|Q| \leqslant 1$, such that

$$J(F, \mathrm{d}m) \int h\mathrm{d}m = \int EhQ\,\mathrm{d}m \qquad (h \in K(\mathrm{d}m),\ hF \in L^1(\mathrm{d}m)).$$

By writing the last relation for $h = f \in A$, $\varphi(f) = 0$ and for $h = 1$, we obtain $QF \in K(\mathrm{d}m)$ and

$$\int QF\mathrm{d}m = J(F, \mathrm{d}m).$$

Since $|QF| \leqslant F$, according to point (2) there results $QF = E$ and, therefore,

$$\int hE\mathrm{d}m = \int h\mathrm{d}m \int E\mathrm{d}m \qquad (h \in K\,\mathrm{d}m,\ hE \in L^1(\mathrm{d}m)).$$

In particular, $hE \in K(\mathrm{d}m)$.

This completes the proof of the proposition. ◇

A function $F \in K(\mathrm{d}m)$ is said to be an inner function if $|F| = 1$, m-almost everywhere.

Theorem 5.12. *Let $h \in K(\mathrm{d}m)$ with*

$$\int h\mathrm{d}m \neq 0.$$

H may then be uniquely written under the form $h = EF$, where E is an outer function and F an inner function. Moreover

$$\int h\mathrm{d}m = \int E\mathrm{d}m \int F\mathrm{d}m.$$

Proof. Jensen inequality (Proposition 5.10) yields

$$J(|h|, \mathrm{d}m) \geqslant |\int h\mathrm{d}m| > 0.$$

According to Proposition 5.11 there exists an outer function $E \in K(\mathrm{d}m)$ such that $|h| = |E|$. The function F defined by $h = EF$ satisfies $|F| = 1$, $F \in L^1(\mathrm{d}m)$ and $EF \in K(\mathrm{d}m)$. Therefore, from the preceding proposition there results that $F \in K(\mathrm{d}m)$ and

$$\int h\mathrm{d}m = \int E\mathrm{d}m \int F\mathrm{d}m. \ \diamond$$

5.4. The characterization of the functions in H^p

We recall that we denoted by $H^p(\mathrm{d}m)$ the $L^p(\mathrm{d}m)$-closure of A for $1 \leqslant p < \infty$. By $H^\infty(\mathrm{d}m)$ we denote the weak $L^\infty(\mathrm{d}m)$-closure of A. Also

$$K(\mathrm{d}m) = \{h \in L^1(\mathrm{d}m) \colon \int fh\mathrm{d}m = \int f\mathrm{d}m \int h\mathrm{d}m,\ f \in A\}.$$

Proposition 5.13. *Let $1 \leqslant p,\ q \leqslant \infty$ be such that $1/p + 1/q = 1$. We have*

(1) $H^p(\mathrm{d}m) \subset K(\mathrm{d}m) \cap L^p(\mathrm{d}m)$,

(2) $H^p(\mathrm{d}m) = \{h \in L^p(\mathrm{d}m) \colon \int hg\mathrm{d}m = 0$ *for any* $g \in K(\mathrm{d}m) \cap L^q(\mathrm{d}m)$ *with* $\int g\mathrm{d}m = 0\}$,

(3) $H^p(\mathrm{d}m) = \{h \in L^p(\mathrm{d}m) \cap K(\mathrm{d}m) \colon \int hg\mathrm{d}m = \int h\mathrm{d}m \int g\mathrm{d}m$ *for any* $g \in L^q(\mathrm{d}m) \cap K(\mathrm{d}m)\}$.

Proof. Since

$$\int fg\mathrm{d}m = \int f\mathrm{d}m \int g\mathrm{d}m \qquad (f, g \in A)$$

there results

$$\int fh\mathrm{d}m = \int f\mathrm{d}m \int g\mathrm{d}m \qquad (f \in A,\ h \in H^p(\mathrm{d}m)),$$

hence point (1) is proved. Relation (2) immediately follows from the duality L^p, L^q and from the $K(\mathrm{d}m)$ definition.

Let $h \in H^p(\mathrm{d}m)$. Then $h \in L^p(\mathrm{d}m) \cap K(\mathrm{d}m)$ according to (1), and from (2) there results

$$\int h(g - \int g\mathrm{d}m)\, \mathrm{d}m = 0$$

for any $g \in L^q(\mathrm{d}m) \cap K(\mathrm{d}m)$, that is

$$\int gh\mathrm{d}m = \int h\mathrm{d}m \int g\mathrm{d}m.$$

Conversely, if $h \in L^p(\mathrm{d}m) \cap K(\mathrm{d}m)$ and

$$\int hg\mathrm{d}m = \int h\mathrm{d}m \int g\mathrm{d}m,$$

for any $g \in L^q(dm) \cap K(dm)$, we then have

$$\int hg dm = 0,$$

for any $g \in K(dm) \cap L^q(dm)$ with

$$\int g dm = 0.$$

From (2) we then get $h \in H^p(dm)$.

Theorem 5.14. *Let $h, g \in K(dm)$ be such that $hg \in L^1(dm)$. Then $hg \in K(dm)$ and we have*

$$\int hg dm = \int h dm \int g dm.$$

Proof. Let

$$T = \{h \in M(X): h \in K(dm) \cap L^\infty(dm)\},$$

$$L: L(h) = \int h dm \qquad (h \in T).$$

T is clearly a linear subspace of $M(X)$, with $1 \in T$, and L a linear functional on T, with $L(1) = 1$. Moreover, using Jensen inequality, we have

$$| L(h) L(g) | = | \int h dm | | \int g dm | \leqslant J(| h |, dm) J(| g |, dm).$$

for $h, g \in K(dm) \cap L^\infty(dm)$. But Lemma 5.1 yields

$$J(| h |, dm) J(| g |, dm) = J(| hg |, dm) \leqslant \int | hg | dm.$$

Therefore,

$$| L(h) L(g) | \leqslant \int | hg | dm$$

and, according to Lemma 5.2, there results

$$L(h) L(g) = \int hg dm,$$

that is

$$\int hg \, dm = \int h dm \int g dm. \tag{5.4.1}$$

Let now $h, g \in K(dm) \cap L^\infty(dm)$. From (5.4.1) there results

$$\int fhg \, dm = \int fh dm \int g dm = \int f dm \int h dm \int g dm = \int f dm \int hg dm$$

for any $f \in A$, hence $hg \in K(dm) \cap L^\infty(dm)$.

Assume now $h, g \in K(\mathrm{d}m)$. Without any loss of generality we may suppose

$$\int h\mathrm{d}m \neq 0, \int g\mathrm{d}m \neq 0.$$

Let then $h = E_1F_1$, $g = E_2F_2$ be the factorizations of h, respectively g, given by Theorem 5.12. Since $F_1, F_2 \in K(\mathrm{d}m) \cap L^\infty(\mathrm{d}m)$, from the first part of the proof we get $F_1F_2 \in K(\mathrm{d}m)$ and

$$\int F_1F_2\mathrm{d}m = \int F_1\mathrm{d}m \int F_2\mathrm{d}m.$$

But E_1 is an outer function in $K(\mathrm{d}m)$ and, therefore, Proposition 5.11 yields $F_1F_2F_1 \in K(\mathrm{d}m)$ and

$$\int F_1F_2E_1\mathrm{d}m = \int F_1F_2\mathrm{d}m \int E_1\mathrm{d}m.$$

At the same time, E_2 is also an outer function in $K(\mathrm{d}m)$ and since $F_1F_2E_1 \in K(\mathrm{d}m)$, the same Proposition 5.11 yields $F_1F_2E_1E_2 \in K(\mathrm{d}m)$ and

$$\int F_1F_2E_1E_2\mathrm{d}m = \int F_1F_2E_1\mathrm{d}m \int E_2\mathrm{d}m.$$

Hence $hg = E_1E_2F_1F_2 \in K(\mathrm{d}m)$ and

$$\int hg\mathrm{d}m = \int h\mathrm{d}m \int g\mathrm{d}m. \diamond$$

Corollary 5.15. *We have*

$$H^p(\mathrm{d}m) = K(\mathrm{d}m) \cap L^p(\mathrm{d}m).$$

Proof. The proof follows from Proposition 5.13 point (3) and Proposition 5.14. ◇

Corollary 5.16. *Any inner function belongs to $H^p(\mathrm{d}m)$, for any $1 \leqslant p \leqslant \infty$. If an outer function belongs to $L^p(\mathrm{d}m)$ it then belongs also to $H^p(\mathrm{d}m)$.*

Corollary 5.17. *Let $h \in H^p(\mathrm{d}m)$ with*

$$\int h\mathrm{d}m \neq 0.$$

Then $h = EF$ where E is an outer function in $H^p(\mathrm{d}m)$ and F an inner function in $H^p(\mathrm{d}m)$.

Corollary 5.18. *Let*

$$A_m = \{f \in A : \int f\,\mathrm{d}m = 0\}.$$

We then have

$$H^p(\mathrm{d}m) = \{h \in L^p(\mathrm{d}m) \colon \int fh\mathrm{d}m = 0, f \in A_m\}.$$

Corollary 5.19. *We have*

$$L^2(\mathrm{d}m) = H^2(\mathrm{d}m) \oplus \overline{H}^2_m,$$

where H^2_m is the L^2(dm)-closure of A_m and $\overline{H}^2_m = \{\bar{h} \colon h \in H^2_m\}$.

Corollary 5.20. *Any real function in H^2 is a constant function.*

To end this paragraph we give the following characterization of the outer functions in $H^p(\mathrm{d}m)$.

Theorem 5.21. *Let $F \in H^p(\mathrm{d}m)$, $1 \leqslant p < \infty$.*

Then $F = cE$, where c is a nonzero constant and E an outer function in $H^p(\mathrm{d}m)$, if and only if the subspace $FH^p(\mathrm{d}m)$ is dense in $H^p(\mathrm{d}m)$.

If E is an outer function in $H^\infty(\mathrm{d}m)$, the subspace $EH^\infty(\mathrm{d}m)$ is then weakly dense in $H^\infty(\mathrm{d}m)$.

Proof. Let E be an outer function in $H^p(\mathrm{d}m)$, $1 \leqslant p \leqslant \infty$, and let $h \in L^q(\mathrm{d}m)$, $1/p + 1/q = 1$, such that

$$\int fEh\,\mathrm{d}m = 0 \qquad (f \in A).$$

Then $Eh \in K(\mathrm{d}m)$ and

$$\int Eh\mathrm{d}m = 0.$$

From Proposition 5.11 there results that $h \in K(\mathrm{d}m)$ and

$$\int h\mathrm{d}m = 0.$$

Therefore,

$$\int fh\mathrm{d}m = 0 \qquad (f \in A),$$

i.e. h is orthogonal to $H^p(\mathrm{d}m)$. The subspace $EH^p(\mathrm{d}m)$ is hence weakly dense in $H^p(\mathrm{d}m)$ for $1 \leqslant p \leqslant \infty$ and, therefore, norm dense for $1 \leqslant p < \infty$.

We obviously reach the same conclusion if we take cE instead of E, where c is a non-zero constant.

Let now F be a function in $H^p(dm)$, $1 \leqslant p \leqslant \infty$, such that the subspace $FH^p(dm)$ is dense in $H^p(dm)$. Since 1 belongs to the $L^p(dm)$-closure of the subspace $FH^p(dm)$, there results

$$c = \int F dm \neq 0.$$

Let $E = \frac{1}{c} F$. We have

$$\int E dm = 1$$

and, according to Jensen inequality, we get

$$1 = \int E dm \leqslant J(|E|, dm).$$

On the other hand, from Szegö theorem there results

$$J(|E|, dm) = D^1(|E|\, dm) =$$

$$= \inf \{ \int |f|\, |E|\, dm, f \in A,\ \varphi(f) = 1\} \leqslant 1$$

since 1 belongs to the L^p -closure of the subspace $\{Ef : f \in A,\ \varphi(f) = 1\}$, hence also to its L^1-closure. Therefore,

$$\int E\, dm = J(|E|, dm) = 1 > 0,$$

i.e. E is an outer function. ◇

5.5. Invariant subspaces

Let A be a function algebra on X, $\mathfrak{M}$ its maximal ideal space, and $\varphi \in \mathfrak{M}$. m will denote a representing measure (on X) for φ with the property that any representing measure for φ, absolutely continuous with respect to m, is identical with m.

A linear closed subspace S of $L^p(dm)$ (norm-closed in $L^p(dm)$ for $1 \leqslant p < \infty$, and weakly closed in $L^\infty(dm)$ if $p = \infty$) will be called *invariant* if for any $f \in A$ and $h \in S$ we have $fh \in S$.

From the duality $L^p(dm) - L^q(dm)$, we obtain that a subspace $S \subset L^p(dm)$ is invariant if and only if it is weakly closed and $AS \subset S$.

If S is an invariant subspace of $L^p(dm)$, we denote by S_5 the weak closure, in $L(dm)^p$, of the set

$$\{hf:\ h \in S,\ f \in A,\ \int f dm = 0\}.$$

Obviously $S_0 \subset S$. The invariant subspace S is said to be *simply invariant* if $S_0 \neq S$.

For a subset $S \subset L^p(\mathrm{d}m)$, we write

$$S^\perp = \{g \in L^p(\mathrm{d}m): \int hg \, \mathrm{d}m = 0, h \in S\}.$$

If S is a simply invariant subspace of $L^p(\mathrm{d}m)$, then it is easy to see 1) that $S_0^\perp$ and $S^\perp$ are invariant subspace in $L^q(\mathrm{d}m)$ and $S^\perp \subset S_0^\perp$, 2) $S^\perp \neq S_0^\perp$.

3) A function $g \in L^q(\mathrm{d}m)$ belongs to $S_0^\perp$ if and only if

$$\int gfh \, \mathrm{d}m = 0$$

for any $h \in S$ and $f \in A$ with $\varphi(f) = 0$, i.e. if and only if $gh \in K(\mathrm{d}m)$ for any $h \in S$. There follows at once that $[S_0^\perp]_0 \subset S^\perp$, hence, if S is a simply invariant subspace of $L^p(\mathrm{d}m)$ then $S_0^\perp$ is a simply invariant subspace of $L^q(\mathrm{d}m)$.

Theorem 5.22. *Let $F \in L^\infty(\mathrm{d}m)$ with $|F| = 1$. The subspace $S = FH^q(\mathrm{d}m)$ of $L^p(\mathrm{d}m)$ is simply invariant.*

Then

$$S_0 = \{Fh, h \in H^p(\mathrm{d}m), \int h\mathrm{d}m = 0\}$$

and

$$S_0^\perp = \bar{F}H^q(\mathrm{d}m).$$

Proof. S is obviously an invariant subspace of $L^p(\mathrm{d}m)$. The set

$$\{Fh: h \in H^p(\mathrm{d}m), \int h\mathrm{d}m = 0\}$$

is weakly closed and, according to the definition of S_0, it contains S_0. Since $F \in S$, there results that for any $f \in A$ with $\varphi(f) = 0$ we have $Ef \in S_0$ and then, as F is bounded and A is dense in $H^p(\mathrm{d}m)$, we obtain $Fh \in S_0$ for any $h \in H^p(\mathrm{d}m)$ with

$$\int h\mathrm{d}m = 0.$$

Hence

$$S_0 = \{Fh: h \in H^p(\mathrm{d}m), \int h\mathrm{d}m = 0\}$$

and, therefore, $S_0 \neq S$, i.e. S is simply invariant.

If $g = Fh$ belongs to S_0, then

$$\int g\bar{F}\,dm = \int Fh\bar{F}\,dm = \int h\,dm = 0$$

and $\bar{F} \in S_0^{\perp}$. Since $S_0^{\perp}$ is invariant, there results that $\bar{F}H^q(dm) \subset S_0^{\perp}$. Let now $g \in S_0$. We then see that for any $h \in S$ we have $gh \in K(dm)$.

Let $h_1 \in K(dm)$ with $Fg = h_1$. Since $F \in L^{\infty}(dm)$ and $g \in L^q(dm)$, there results that $h_1 \in L^q(dm)$ and since $h_1 \in K(dm)$, we have $h_1 \in H^q(dm)$, hence $g = Fh_1$ with $h_1 \in H^q(dm)$, that is $S_0^{\perp} = \bar{F}H^q(dm)$.

This completes the proof. ◇

Theorem 5.23. *Let S be a simply invariant subspace of $L^2(dm)$. Then there exists a function $F \in S$ such that $|F| = 1$ and $S = FH^2(dm)$. The function F is uniquely determined up to a constant factor of modulus* 1.

Proof. Let $F \in S$ be orthogonal on S_0 and

$$\int |F|^2 dm = 1.$$

Since $fF \in S_0$ for any $f \in A$ with $\varphi(f) = 0$, there results

$$\int f|F|^2 dm = 0 \qquad (f \in A,\ \varphi(f) = 0).$$

From Corollary 5.18 there results $|F|^2 \in H^2(dm)$ and, since it is real, according to Corollary 5.20, we have $|F| = 1$. Therefore, the multiplication by F is an isometry in $L^2(dm)$ and, since $F \in S$ and S is invariant, we obtain $FH^2(dm) \subset S$.

Let now $h \in S$ orthogonal on $FH^2(dm)$. Then, for any $f \in A$ we have

$$\int fF\bar{h}\,dm = 0 \qquad (f \in A),$$

and, in particular, $F\bar{h} \in H^2(dm)$. On the other hand, since F is orthogonal on S_0, we get

$$\int fh\,\bar{F}\,dm = 0 \qquad (f \in A,\ \varphi(f) = 0)$$

that is $h\bar{F} \in H^2(dm)$. Then, according to Corollary 5.20, $F\bar{h}$ is constant and, since

$$\int \bar{h}\,F\,dm = 0,$$

we have $F\bar{h} = 0$. But $|F| = 1$, hence $h = 0$ and, therefore, $S = FH^2(dm)$.

If $S = F_1H^2(dm) = F_2H^2(dm)$, then $F_1 = F_2h_2$, $F_2 = F_1h_1$ with $h_1, h_2 \in H^2(dm)$. There results that $F_1\bar{F}_2 \in H^2(dm)$, $\bar{F}_1F_2 \in H^2(dm)$ and from Corollary 5.20, $F_1\bar{F}_2$ is constant. Therefore F_1 differs from F_2 by a constant factor, whose modulus is obviously 1.

Theorem 5.24. *Let F be a function in $L^\infty(dm)$ with $|F| = 1$. The subspace $FH^p(dm)$ is simply invariant in $L^p(dm)$ for any p, $1 \leqslant p < \infty$. Conversely, if S is a simply invariant subspace of $L^p(dm)$, then there exists a function $F \in L^\infty(dm)$, $|F| = 1$, uniquely determined up to a constant factor of modulus 1, such that $S = FH^p(dm)$.*

Proof. The first part of the theorem follows from Theorem 5.22.

Let now S be a simply invariant subspace of $L^p(dm)$. We first assume $1 \leqslant p < 2$. Let $T = S \cap L^2(dm)$. T is obviously an invariant subspace of $L^2(dm)$. Considered as a subspace of S, T is dense in S. Indeed, let $h \in S$ and F_n be the function of $L^\infty(dm)$ defined by

$$F_n = \begin{cases} 1 & \text{if } |f| \leqslant n \\ |f|^{-1} & \text{if } |f| > n. \end{cases}$$

We clearly have $J(F_n, dm) > 0$ and, according to Proposition 5.11, there exists an outer function E_n in $H^\infty(dm)$, with $|E_n| = F_n$. Since the sequence E_n is bounded in norm, in $L^1(dm)$, it contains a subsequence, also denoted by E_n, convergent almost everywhere in $L^1(dm)$. Let E be its limit in $L^1(dm)$; we have

$$\int E\, dm = \lim \int E_n\, dm = \lim \ J(|E_n|, dm) = J(|E|, dm).$$

On the other hand, since clearly $|E_n| = 1$, there results $|E| = 1$. Therefore, E is an outer function, and $E = 1$ follows from $|E| = 1$ and the uniqueness part of Proposition 5.11.

Since S is invariant and $E_n \in H^\infty(dm)$, we get $E_nf \in S$ and since $|E_nf| \leqslant n$, we have $E_nf \in T$. On the other hand, it is clear that the sequence E_nf boundedly converges to f and, therefore, it converges to f in $L^p(dm)$.

We thus proved that T is $L^p(dm)$-dense in S. From this fact and the simple invariance of S, T results simply invariant.

According to Theorem 5.22, there exists a function $F \in T$ with $|F| = 1$ and $T = FH^2(dm)$. Since T is dense in S, there results $S = FH^p(dm)$.

Let now S be a simply invariant subspace in $L^p(dm)$, $p > 2$, and let $q < 2$ with $1/p + 1/q = 1$.

Consider $T = S_0$. We then know that T is a simply invariant subspace in $L^q(dm)$ and $T_0 \subset S^\perp \subset T$, $S^\perp \neq T$. According to the first part of the proof and to Theorem 5.22, there is an $F \in T$ such that $|F| = 1$ and $T = FH^q(dm)$, and

$$T_0 = \{Fh\colon h \in H^q(dm), \int h dm = 0\}.$$

Let $g \in S^\perp$, $g = Fh_1$ with $h_1 \in H^q(dm)$.
There follows

$$\int Fhh_1 dm = 0 \qquad (h \in S).$$

Since $F \in T = S_0^\perp$ and $h \in S$, we know that $Fh \in K(dm)$.
By applying the multiplicativity theorem (Theorem 5.14) we obtain

$$\int Fh\, dm \int h_1 dm = 0, \qquad (h \in S).$$

If

$$\int Fh\, dm = 0 \qquad (h \in S)$$

there results that $F \in S^\perp$, hence $T = FH^p(dm) \subset S^\perp$ which is impossible. Therefore,

$$\int h_1\, dm = 0$$

and we get $g = Fh_1 \in T_0$. There follows that $S^\perp = T_0$ and by Theorem 5.22 we obtain

$$S = T_0^\perp = \bar{F}H^p(dm).$$

Suppose now that for $S \subset L^p(dm)$, $1 \leqslant p \leqslant \infty$, there exist F_1, $F_2 \in S$ such that $|F_1| = |F_2| = 1$ and $S = F_1H^p(dm) = F_2H\,(dm)$. Then $F_1 = F_2h_2$, $F_2 = F_1h_1$ with h_1, $h_2 \in H^p(dm)$. There results that $F_1\bar{F}_2 = h_2 \in H^p(dm)$, $F_2\bar{F}_1 = h_1 \in H^p(dm)$, hence $F_1\bar{F}_2$, $F_2\bar{F}_1 \in H^2(dm)$. Then, according to Corollary 5.20, F_1 and F_2 differ by a constant factor of modulus 1.

The theorem is proved. ◇

Theorem 5.25. *Let* $g \in L^p(\mathrm{d}m)$, $1 \leqslant p \leqslant \infty$ *and* S *be the invariant subspace generated by* g *in* $L^p(\mathrm{d}m)$.

S is a simply invariant subspace if and anly if $J(|f|, \mathrm{d}m) > 0$.

Proof. If $J(|g|, \mathrm{d}m) > 0$ then, by Theorem 5.11, there exists an outer function $E \in H^p(\mathrm{d}m)$ such that $|E| = g$. Then clearly $g = FE$ with $F \in L^\infty(\mathrm{d}m)$, $|F| = 1$. Since the space S is equal to the weak closure in $L^p(\mathrm{d}m)$ of the set $\{gf, f \in A\}$, from Theorem 5.21 we get $S = FH^p(\mathrm{d}m)$ and from Theorem 5.24, S results simply invariant.

Assume now that S is simply invariant. According to Theorem 5.24 there exists $|F| = 1$ such that $S \subset FH^p(\mathrm{d}m)$. Let $g = Fh$ with $h \in H^p(\mathrm{d}m)$. Since the set $\{gf, f \in A\}$ is weakly dense in S and $F \in S$, there exists $f \in A$ such that

$$|\int (fg - F)\bar{F}\,\mathrm{d}m| < 1.$$

We then have

$$|\int f\mathrm{d}m \int h\mathrm{d}m - 1| = |\int (fh - 1)\mathrm{d}m| = |\int (fg - F)\bar{F}\mathrm{d}m| < 1.$$

There results

$$\int h\mathrm{d}m \neq 0$$

and by Jensen inequality we get

$$J(|g|, \mathrm{d}m) = J(|h|, \mathrm{d}m) \geqslant |\int h\mathrm{d}m| > 0. \quad \diamond$$

5.6. The algebra H^∞ (dm)

Let A be a function algebra on X and φ a nonzero multiplicative linear functional on A. Let m be a representing measure for φ with the property that any representing measure for φ, absolutely continuous with respect to m, is identical with m.

The algebra $L^\infty(\mathrm{d}m)$ is a commutative Banach algebra, with unit element, whose norm satisfies

$$\|f^2\| = \|f\|^2 \qquad (f \in L^\infty(\mathrm{d}m)).$$

If we denote by Y the maximal ideal space of $L^\infty(\mathrm{d}m)$, then it is known that $L^\infty(\mathrm{d}m)$ is isometrically embedded into $C(Y)$. Since $L^\infty(\mathrm{d}m)$ is obviously symmetric, from the Stone-Weierstrass Theorem we get $L^\infty(\mathrm{d}m) = C(Y)$.

The algebra $H^\infty(\mathrm{d}m)$ is a subalgebra of $L^\infty(\mathrm{d}m)$, hence a subalgebra of $C(Y)$, which contains the constant functions and is uniformly closed.

According to Theorem 5.11, $H^\infty(\mathrm{d}m)$ separates the points of Y. $H^\infty(\mathrm{d}m)$ is therefore, a function algebra on Y.

Theorem 5.26. *For any real function* $u \in L^\infty(\mathrm{d}m)$ *there exists* $E \in H^\infty(\mathrm{d}m)$ *such that* $E^{-1} \in H^\infty(\mathrm{d}m)$ *and* $e^u = |E|$.

Proof. The functions $F = e^u$ and $F_1 = e^{-u}$ belong to $L^\infty(\mathrm{d}m)$ and we obviously have $J(F, \mathrm{d}m) > 0$, $J(F_1, \mathrm{d}m) > 0$. According to Theorem 5.11 there exist the outer functions E and E_1 in $H^\infty(\mathrm{d}m)$ such that $|E| = F$, $|E_1| = F_1$. Since EE_1 is an outer function and $|EE_1| = 1$, there results $EE_1 = 1$ and, therefore, $E^{-1} \in H^\infty(\mathrm{d}m)$. ◇

Notes

The results contained in this chapter have their origin in classical problems of the theory of analytic functions of a complex variable and especially in the theory of Hardy H^p-classes. A modern treatment of all these classical results can be found in K. Hoffman's work [2].

H. Helson and D. Lowdenslager [1], [2] were the first to present a theory of H^p-spaces under more general conditions than Hardy's theory. There followed a series of works which proved in more and more general conditions, results similar to classical ones. We mention here some of them: P. R. Ahern, D. Sarason [1], I. Glicksberg [3], K. Hoffman [1], K. Hoffman, H. Rossi [1], H. König [1], [2], [3], [4], G. Lumer [1], [2], etc.

The present chapter follows mainly the synthesis work of H. König [4].

CHAPTER 6

Special classes of function algebras

6.1. Dirichlet and logmodular algebras

Let A be a function algebra on X. A is called a *Dirichlet algebra on* X if the space $\mathrm{Re}A$ is (uniformly) dense in $C_R(X)$. A is called a *logmodular algebra* on X if the set $\log|A^{-1}| = \{u \in C_R(X)\colon u = \log|f|, f \in A, f^{-1} \in A\}$ is dense in $C_R(X)$.

Since for any $f \in A$ we have $e^f \in A^{-1}$ and $\mathrm{Re}f = \log|e^f|$, there results $\mathrm{Re}A \subset \log|A^{-1}|$ and therefore *any Dirichlet algebra on* X *is a logmodular algebra on* X.

Theorem 6.1. *Let* A *be a logmodular algebra on* X, $\mathfrak{M}$ *its maximal ideal space and* $\varphi \in \mathfrak{M}$. *There exists only one representing measure for* φ *with support in* X.

Proof. Let μ_1, μ_2 be two representing measures for φ, with support in X. If $f \in A^{-1}$, we have

$$f(\varphi) = \int f d\mu_1, \qquad |f(\varphi)| \leqslant \int |f| \,\mathrm{d}\mu_1$$

$$f^{-1}(\varphi) = \int f^{-1} \,\mathrm{d}\mu_2, \qquad |f^{-1}(\varphi)| \leqslant \int |f|^{-1} \,\mathrm{d}\mu_2.$$

Since $f(\varphi)f^{-1}(\varphi) = 1$, there results

$$1 \leqslant \int |f| \,\mathrm{d}\mu_1 \int |f|^{-1} \,\mathrm{d}\mu_2 \qquad (f \in A^{-1}).$$

But the set $\log|A^{-1}|$ is dense in $C_R(X)$, therefore for any function e^u with $u \in C_R(X)$ we have

$$1 \leqslant \int e^u \mathrm{d}\mu_1 \int e^{-u} \mathrm{d}\mu_2. \tag{6.1.1}$$

We now fix $u \in C_R(X)$ and, for any real t, consider

$$\rho(t) = \int e^{tu} d\mu_1 \int e^{-tu} d\mu_2.$$

$\rho(t)$ is a differentiable function and from (6.11) we get $\rho(t) \geqslant 1$ for any t. Since $\rho(0) = 1$, there results $\rho'(0) = 0$. Then

$$0 = \rho'(0) = \int u d\mu_1 - \int u d\mu_2$$

hence

$$\int u d\mu_1 = \int u d\mu_2$$

for any $u \in C_R(X)$ and, therefore, $\mu_1 = \mu_2$. ◇

Corollary 6.2. *For any $x \in X$, ε_x is the only representing measure for x. Then, the Choquet boundary of A is equal to its Shilov boundary and both are identical with X.*

Due to the uniqueness of the representing measure for the elements φ of $\mathfrak{M}$, the results obtained in the previous Chapter on H^p-spaces hold for any $\varphi \in \mathfrak{M}$ and its representing measure m on X.

As an application of these results, we shall determine the structure of the Gleason parts of the maximal ideal space of a logmodular algebra A on X. It is worth noting that the proofs hold also for a more general case, when any element $\varphi \in \mathfrak{M}$ admits a unique representing measure with support in X.

An element $\varphi_1 \in \mathfrak{M}$ is called *bounded* in $H^2(dm)$ if there exists a constant c such that

$$|\varphi_1(f)| \leqslant c[\int |f|^2 dm]^{1/2} \qquad (f \in A).$$

A functional in $\mathfrak{M}$ which is bounded in $H^2(dm)$ can be continuously extended to a linear multiplicative functional on $H^2(dm)$.

Let us denote

$$H^2_m = \{h \in H^2(dm): \int h dm = 0\}.$$

The subspace H^2_m of $H^2(dm)$ is obviously invariant.

Proposition 6.3. *Assume the subspace H^2_m simply invariant and let $H^2_m = ZH^2(\mathrm{d}m)$ be its writing given by Theorem 5.23. For $h \in H^2(\mathrm{d}m)$ we write*

$$a_n = \int \bar{Z}^n h \, \mathrm{d}m \qquad (n = 0, 1, 2, \ldots).$$

For any $\varphi_1 \in \mathfrak{M}$, bounded in $H^2(\mathrm{d}m)$, and $h \in H^2(\mathrm{d}m)$ we have $|\varphi_1(Z)| < 1$ and

$$\varphi_1(h) = \sum_{n=0}^{\infty} a_n[\varphi_1(Z)]^n.$$

The measure

$$\mathrm{d}m_1 = \frac{1 - |\varphi_1(Z)|^2}{|Z - \varphi_1(Z)|^2} \mathrm{d}m$$

is the representing measure of φ_1.

Proof. We first show that $|\varphi_1(Z)| < 1$. Indeed, since φ_1 is bounded in $H^2(\mathrm{d}m)$, there results

$$|\varphi_1(Z)|^n = |\varphi_1(Z^n)| < C[\int |Z^n|^2 \, \mathrm{d}m]^{1/2} = C$$

for any natural n and therefore $|\varphi_1(Z)| \leqslant 1$.

We suppose $|\varphi_1(Z)| = 1$. Since Z is determined, but a constant factor of modulus 1, we can take $\varphi_1(Z) = 1$. For a natural n let $f_n = \sum_{k=1}^{\infty} b_k z^k$ be a function of the standard algebra such that $f_n(1) = n$ and $\sum_{k=1}^{\infty} |b_k|^2 \leqslant 1$ *). Then the function $h_n = \sum_{k=1}^{\infty} b_k Z^k$ belongs to $H^2(\mathrm{d}m)$ and we have

$$\varphi_1(h_n) = n, \qquad \int |h_n|^2 \, \mathrm{d}m \leqslant 1,$$

which contradicts the fact that φ_1 is bounded.

Therefore $|\varphi_1(Z)| < 1$.

*) Such a function exists since 1 is a Choquet point for the standard algebra.

The sequence a_n is bounded, hence $\sum_{n=0}^{\infty} a_n[\varphi_1(Z)]^n$ is a convergent series.

Let now $h \in H^2(\mathrm{d}m)$. Since $h - \varphi(h) \in H_m^2$ we get $h - \varphi(h) = Zh_1$ with $h_1 \in H^2(\mathrm{d}m)$. Hence $\bar{Z}[h - \varphi(h)]$ belongs to $H^2(\mathrm{d}m)$ and we have

$$\int |\bar{Z}[h - \varphi(h)]|^2 \,\mathrm{d}m = \int |h - \varphi(h)|^2 \,\mathrm{d}m = \int |Zh_1|^2 \,\mathrm{d}m \leqslant \int |h|^2 \mathrm{d}m$$

as Zh_1 is the projection of h on H_m^2, the constants being orthogonal on $ZH^2(\mathrm{d}m)$.

Thus, the linear operator T

$$Th = \bar{Z}(h - \int h \,\mathrm{d}m) \qquad (h \in H^2(\mathrm{d}m))$$

defined on $H^2(\mathrm{d}m)$ with values in $H^2(\mathrm{d}m)$, is bounded and $\|T\| \leqslant 1$. We have

$$\int Th\mathrm{d}m = \int \bar{Z}h\mathrm{d}m - \int h\mathrm{d}m \int \bar{Z}\mathrm{d}m = \int \bar{Z}h\mathrm{d}m = a_1.$$

Furthermore

$$T^2h = T(Th) = \bar{Z}(Th - \int T h\mathrm{d}m) = \bar{Z}(Th - a_1),$$

hence

$$Th = a_1 + Z[T^2h].$$

Therefore

$$h = \int h\mathrm{d}m + ZTh = a_0 + a_1Z + Z^2[T^2h].$$

Let us assume

$$h = a_0 + a_1Z + \dots + a_{n-1}Z^{n-1} + Z^n[T^nh].$$

We then clearly have

$$\int T^nh \,\mathrm{d}m = \int \bar{Z}^n h\mathrm{d}m = a_n$$

which implies

$$T^{n+1}h = T(T^nh) = \bar{Z}(T^nh - \int T^nh\mathrm{d}m) = \bar{Z}(T^nh - a_n),$$

that is

$$T^n h = a_n + Z[T^{n+1}h].$$

Then

$$h = a_0 + a_1 Z + \dots + a_n Z^n + Z^{n+1}[T^{n+1}h]$$

and, for any natural number n, we have

$$h = a_0 + a_1 Z + \dots + a_{n-1} Z^{n-1} + Z^n[T^n h].$$

Therefore we obtain

$$\sum_{k=1}^{n} a_k[\varphi_1(Z)]^k = \varphi_1(h) - [\varphi_1(Z)]^{n+1}[\varphi_1(T^{n+1}h)].$$

But $|\varphi_1(Z)| < 1$ and the sequence $\varphi_1(T^{n+1}h)$ is bounded, since φ_1 is bounded in $H^2(\mathrm{d}m)$ and $\|T\| \leqslant 1$; hence

$$\varphi_1(h) = \sum_{n=0}^{\infty} a_n[\varphi_1(Z)]^n \qquad (h \in H^2(\mathrm{d}m)).$$

Let

$$\mathrm{d}m_1 = \frac{1 - |\varphi_1(Z)|^2}{|Z - \varphi_1(Z)|^2}\,\mathrm{d}m.$$

Since

$$\frac{1 - |\varphi_1(Z)|^2}{|Z - \varphi_1(Z)|^2} = \frac{Z}{Z - \varphi_1(Z)} + \frac{\overline{\varphi_1(Z)}}{\bar{Z} - \overline{\varphi_1(Z)}} =$$

$$= \frac{1}{1 - \bar{Z}\varphi_1(Z)} + \frac{\overline{\varphi_1(Z)}Z}{1 - \overline{\varphi_1(Z)}Z} = \sum_{n=0}^{\infty} [\varphi_1(Z)]^n Z^{-n} +$$

$$+ \overline{\varphi_1(Z)}Z \sum_{n=0}^{\infty} Z^n[\overline{\varphi_1(Z)}]^n$$

there results

$$\int f dm_1 = \sum_{n=0}^{\infty} [\varphi_1(Z)]^n \int \bar{Z}^n f dm + \sum_{n=0}^{\infty} \overline{[\varphi_1(Z)]}^{n+1} \int Z^{n+1} f dm =$$

$$= \sum_{n=0}^{\infty} a_n [\varphi_1(Z)]^n = f(\varphi_1)$$

for any $f \in A$, i.e. m_1 is a representing measure for φ_1.
The proposition is completely proved. ◇

Theorem 6.4. *Let* Δ *be the Gleason part of* $\mathfrak{M}$ *which contains* φ. *We have* $\Delta \neq \{\varphi\}$ *if and only if the subspace* H^2_m *is simply invariant. If* $\Delta \neq \{\varphi\}$, *there exists an injective map* τ *of the unit disk* $D = \{z : |z| < 1\}$ *of the complex plane, into* $\mathfrak{M}$, *such that* τ *is continuous and:*

(a) $\tau(D) = \Delta$

(b) *For any* $f \in A$, $F(z) = f[\tau(z)]$ *is an analytic function in* D.

Proof. Assume $\Delta \neq \{\varphi\}$ and let $\varphi_1 \in \Delta$, $\varphi_1 \neq \varphi$ and m_1 be its representing measure. According to Theorem 3.13, there exists c, $0 < c < 1$ such that $cm_1 \leqslant m$, $cm \leqslant m_1$. Hence, the identity map of A into A generates a topological isomorphism between $H^2(dm)$ and $H^2(dm_1)$ and φ and φ_1 are bounded in $H^2(dm_1)$ respectively $H^2(dm)$.

Let S be the $H^2(dm_1)$-closure of the set

$$A_m = \{f \in A : \int f dm = 0\}.$$

Since the $H^2(dm)$-closure of A_m is identical with H^2_m, the identity map of A_m onto A_m generates a topological isomorphism between H^2_m and S.

S is obviously an invariant subspace in $H^2(dm_1)$.

We prove it is simply invariant. Indeed, if $S_0 = S$, then since S_0 is the closure of $\{fh : h \in S, f \in A, \varphi(f) = 0\}$ there results

$$\int h dm_1 = 0 \qquad (h \in S).$$

In particular

$$\varphi_1(f) = \int f dm_1 = 0, \qquad (f \in A_m)$$

hence $\varphi_1 = \varphi$. But $\varphi_1 \neq \varphi$, therefore S is simply invariant in $H^2(dm_1)$. Let $F \in S$ with $|F| = 1$ m_1-almost everywhere and $S = FH^2(dm_1)$. Then, clearly, $|F| = 1$ m-almost everywhere and $H_m^2 = FH^2(dm)$. Then from Theorem 5.22, H_m^2 is simply invariant.

Consider now H_m^2 simply invariant and let $H_m^2 = ZH^2(dm)$ be its form given by Theorem 5.23. If $\varphi_1 \in \Delta$, according to Theorem 3.12, φ_1 is bounded in $H^2(dm)$ and from Proposition 6.3 we get $|\varphi_1(Z)| < 1$.

We define on Δ a mapping θ, by

$$\theta(\varphi_1) = \varphi_1(Z), \qquad (\varphi_1 \in \Delta).$$

We have $\theta(\Delta) \subset D$. If $\theta(\varphi_1) = \theta(\varphi_2)$ for $\varphi_1, \varphi_2 \in \Delta$, from Proposition 6.3 there results

$$\varphi_1(f) = \sum_{n=0}^{\infty} a_n[\varphi_1(Z)]^n = \sum_{n=0}^{\infty} a_n[\varphi_2(Z)]^n = \varphi_2(f)$$

for any $f \in A$, and since A separates $\mathfrak{M}$ we obtain $\varphi_1 = \varphi_2$. Therefore, θ is an injection of Δ into D. Let us prove that θ maps Δ onto D. Let $f, g \in A$ and

$$a_n = \int \bar{Z}^n f dm, \; b_n = \int \bar{Z}^n g dm.$$

We have seen in the proof of Proposition 6.3 that

$$f = a_0 + a_1 Z + \ldots + a_n Z^n + h_1 Z^{n+1}$$

$$g = b_0 + b_1 Z + \ldots + b_n Z^n + h_2 Z^{n+1}$$

where $h_1 = T^{n+1}f$, $h_2 = T^{n+1}g$ and T is the operator defined on $H^2(dm)$ with values in $H^2(dm)$, given by

$$Th = Z(h - \int h dm) \qquad (h \in H^2(dm)).$$

Hence $h_1, h_2 \in H^2(dm)$ and are both bounded.

Let

$$c_n = \int \bar{Z}^n fg dm \qquad (n = 0, 1, 2, \ldots).$$

Using the above expressions of f and g we find by a simple calculation that

$$c_0 = a_0 b_0,$$

$$c_1 = a_0 b_1 + a_1 b_0$$

$$c_2 = a_0 b_2 + a_1 b_1 + a_2 b_0$$

etc.

Let now $\lambda \in D$. For $f \in A$ we define

$$\varphi_1(f) = \sum_{n=0}^{\infty} a_n \lambda^n,$$

where

$$a_n = \int \bar{Z}^n f \mathrm{d}m.$$

Since a_n is a bounded sequence, φ_1 is well defined on A, linear and, according to the above considerations, also multiplicative.

As $\varphi_1(1) = 1$ we have $\varphi_1 \neq 0$. Therefore $\varphi_1 \in \mathfrak{M}$.

Let $f \in A$ with

$$\int |f|^2 \mathrm{d}m < \varepsilon^2.$$

Then $|a_n| < \varepsilon$ for any n, which implies

$$|\varphi_1(f)| < \frac{\varepsilon}{1 - |\lambda|}.$$

Hence φ_1 is bounded in $H^2(\mathrm{d}m)$. The representing measure of φ_1 is, according to Proposition 6.3,

$$\mathrm{d}m_1 = \frac{1 - [\varphi_1(Z)]^2}{|Z - \varphi_1(Z)|^2} \mathrm{d}m.$$

Therefore m and m_1 are mutually absolutely continuous, with bounded derivatives and, from Theorem 3.12, $\varphi_1 \in \Delta$.

Since, obviously, $\theta(\varphi_1) = \lambda$, θ is an injection of Δ onto D, hence Δ does not reduce to only one point.

Let now $\tau = \theta^{-1}$ be the inverse map of θ. τ is a one-to-one correspondence between D and Δ and, for any $f \in A$, we have

$$F(\lambda) = f[\tau(\lambda)] = \sum_{n=0}^{\infty} a_n \lambda^n,$$

where a_n is the bounded sequence

$$a_n = \int \bar{Z}^n f^n \mathrm{d}m.$$

Therefore F is an analytic function in D and τ is continuous.

The theorem is thus proved. ◇

To conclude this paragraph we give a variant to F. and M. Riesz theorem in which the Lebesgue decomposition with respect to a representing measure is replaced with a decomposition with respect to all representing measures; the sense of this decomposition will be specified further.

We first prove the following completion to Theorem 3.12.

Proposition 6.5. *Let $\varphi_1, \varphi_2 \in \mathfrak{M}$ such that φ_1 and φ_2 do not belong to the same Gleason part of $\mathfrak{M}$. Then, the representing measures μ_1, μ_2 of φ_1, respectively φ_2 are mutually singular.*

Proof. Let $\mathrm{d}\mu_2 = h\mathrm{d}\mu_1 + \mathrm{d}\mu_s$ be the Lebesgue decomposition of μ_2 with respect to μ_1. Since μ_1, μ_2 are positive measures of norm 1, there results $0 \leqslant h \leqslant 1$. We then have

$$2 = \|\varphi_1 - \varphi_2\| \leqslant \|\mu_1 - \mu_2\| = \|(1 - h)\mu_1 + \mu_s\| =$$

$$= \|(1 - h)\mu_1\| + \|\mu_s\| \leqslant \int (1 - h)\, \mathrm{d}\mu_1 + 1 = 2 - \int h\mathrm{d}\mu_1.$$

Hence

$$\int h\mathrm{d}\mu_1 \leqslant 0$$

and, since h is positive, we obtain $h = 0$ μ_1-almost everywhere, which implies $\mu_2 = \mu_s$. ◇

Theorem 6.6. *Let A be a logmodular algebra on X and $\mu \in A^{\perp}$. There exist: an at most countable set $\{\varphi_i\}$ of distinct elements in $\mathfrak{M}$ any two elements belonging to two distinct Gleason parts, the func-*

tions $h_i \in H^1_{\lambda_i}$ *where* λ_i *are the representing measures, for* φ_i, *and the measure* $\sigma \in A^\perp$, *singular with respect to all representing measures for the elements of* $\mathfrak{M}$, *such that*

$$\mu = \sum_{i=1}^{\infty} h_i \lambda_i + \sigma,$$

the series being convergent in norm.

Proof. If μ and λ are two measures on X, we denote by μ_λ the absolutely continuous part of μ relative to λ, and by μ'_λ the singular part of μ with respect to λ.

Let $\varphi_1, \ldots, \varphi_n$ be a finite system of elements in $\mathfrak{M}$, each two elements belonging to two distinct Gleason parts and $\lambda_1, \ldots, \lambda_n$ their representing measures. According to Proposition 6.5, λ_i and λ_j are mutually singular for $i \neq j$. Let $\mu \in A^\perp$ and

$$\rho = \mu - \sum_{i=1}^{n} \mu_{\lambda_i}.$$

We have

$$\rho = \mu'_{\lambda_j} - \sum_{i \neq j} \mu_{\lambda_i}, \qquad (j = 1, 2, \ldots, n)$$

hence ρ is singular with respect to every λ_i.

Since

$$\mu = \rho + \sum_{i=1}^{n} \mu_{\lambda_i}$$

there results

$$\|\mu\| = \|\rho\| + \sum_{i=1}^{n} \|\mu_{\lambda_i}\|$$

which implies

$$\sum_{i=1}^{n} \|\mu_{\lambda_i}\| \leqslant \|\mu\|.$$

This inequality holds for any system $\lambda_1,\ldots,\lambda_n$ with the above properties, therefore there exists at most a countable set of Gleason parts which contain an element φ with $\mu_\lambda \neq 0$, λ being the representing measure of φ.

Let $\{\Delta_i\}$ be this set. We fix $\varphi_i \in \Delta_i$ and let λ_i be the representing measure of φ_i. Since

$$\sum_{i=1}^{\infty} \|\mu_{\lambda_i}\| \leqslant \|\mu\|$$

the series $\sum_{i=1}^{\infty} \mu_{\lambda_i}$ is convergent in norm. Let

$$\sigma = \mu - \sum_{i=1}^{\infty} \mu_{\lambda_i}.$$

Let λ be the representing measure of an element $\varphi \in \mathfrak{M}$. If there is a j such that $\varphi \in \Delta$, then λ and λ_j are mutually absolutely continuous and, for $i \neq j$, λ and λ_i are mutually singular. There results $(\mu_{\lambda_j})_\lambda = \mu_\lambda$ and $(\mu_{\lambda_i})_\lambda = 0$ for $i \neq j$, hence $\sigma_\lambda = \mu_\lambda - \mu_\lambda = 0$. If $\varphi \in \Delta_i$ for any i, then, from the properties of the set $\{\Delta_i\}$, we get $\mu_\lambda = 0$ and, according to Proposition 6.5, λ and λ_i are mutually singular for any i, therefore $(\mu_{\lambda_i})_\lambda = 0$ which implies $\sigma_\lambda = 0$. Thus σ is singular with respect to any representing measure of the elements of $\mathfrak{M}$.

Following Theorem 5.6 we have $\mu_{\lambda_i} \in A^\perp$ for any i, hence $\sigma \in A^\perp$.

Let $h_i \in L^1(\mathrm{d}\lambda_i)$ with $\mathrm{d}\mu_{\lambda_i} = h_i \mathrm{d}\lambda_i$. Since $\mu_{\lambda_i} \in A^\perp$, there results

$$\int f h_i \mathrm{d}\lambda_i = 0 \qquad (f \in A)$$

and, from Corollary 5.18, we get $h_i \in H^1_{\lambda_i}$.

This completes the proof of the theorem. $\diamond$

The importance of this theorem consists in the following: there are function algebras A with the property that the only measure orthogonal to A and singular with respect to any multiplicative measure on A, is the null measure; at the same time, there are orthogonal measures to A which are absolutely continuous with respect to no representing measure. In this sense give an example in § 6.4.

6.2. Algebras generated by inner functions

Let A be a function algebra on X, $\mathfrak{M}$ its maximal ideal space and Γ its Shilov boundary.

Let

$$S_1 = \{s \in A,\ s \neq 0\colon \|fs\| = \|f\| \text{ for any } f \in A\}.$$

We obviously have $\|s\| = 1$ for any $s \in S_1$.

Proposition 6.7. *The function s of A belongs to S_1 if and only if $|s| = 1$ on the Shilov boundary, i.e.*

$$S_1 = \{s \in A\colon |s(x)| = 1,\ x \in \Gamma\}.$$

Proof. If $|s(x)| = 1$ for any $x \in \Gamma$, then $\|s\| = 1$ and

$$\|sf\| = \sup_{x \in \Gamma} |s(x)f(x)| = \sup_{x \in \Gamma} |s(x)|\,|f(x)| = \sup_{x \in \Gamma} \|f(x)\| = \|f\|$$

for any $f \in A$, that is $s \in S_1$.

Conversely, assume there exist $s \in S_1$ and $x_0 \in \Gamma$ such that $|s(x_0)| < 1$. Let $\varepsilon > 0$ and U be the neighbourhood of x_0 on which

$$|s(x)| < 1 - \varepsilon \qquad (x \in U).$$

According to Theorem 2.16 there exists $f \in A$ with $\|f\| = 1$ and $|f| < \varepsilon$ off U. We then have

$$|s(x)f(x)| = |s(x)|\,|f(x)| < 1 - \varepsilon \qquad (x \in U)$$

and

$$|s(x)f(x)| = |s(x)|\,|f(x)| < \varepsilon,$$

that is $\|sf\| < 1$, which contradicts the relation $\|sf\| = \|f\| = 1$. ◇

Corollary 6.8. *If $\Gamma = \mathfrak{M}$ then any element of S_1 is invertible.*

Corollary 6.9. *If $s \in S_1$ and is invertible then $|s| = 1$ on $\mathfrak{M}$.*

Proof. Since $\|s\|\ \|s^{-1}\| = \|ss^{-1}\| = 1$, we get $\|s^{-1}\| = 1$. Then for any $\varphi \in \mathfrak{M}$ we have

$$|s(\varphi)| = \frac{1}{|s^{-1}(\varphi)|} \geqslant 1$$

and, since $\|s\| = 1$, there results $|s(\varphi)| = 1$.

Let now S be a multiplicatively closed system of functions in S_1. Assume $1 \in S$. The $C(\Gamma)$-closure A_s of the set $\{g \in C(\Gamma): g = \bar{s}f,\ s \in S,\ f \in A\}$ is a function algebra on Γ. Indeed, since S is multiplicatively closed and $|s| = 1$ on Γ for any $s \in S$, there results, on Γ,

$$\bar{s}_1 f_1 + \bar{s}_2 f_2 = \overline{s_1 s_2}(s_2 f_1 + s_1 f_2)$$

$$\bar{s}_1 f_1 \bar{s}_2 f_1 = \bar{s}_1 \bar{s}_2 f_1 f_2,$$

hence A_s is a uniformly closed subalgebra of $C(\Gamma)$. Obviously $A \subset A_s$, therefore A_s separates the points of Γ and contains the constants.

As $\bar{S} \in A_s$, any $s \in S$ is invertible in A_s. We call A_s the algebra of quotients of A with numerators in S.

Let $\mathfrak{M}_s$ be the maximal ideal space of A_s and Γ_s its Shilov boundary. The embeddings defined in paragraph 3.1 allow us to write $\Gamma_s \subset \Gamma \subset \subset \mathfrak{M}_s \subset \mathfrak{M}$; indeed, we easily verify that A, as subalgebra in A_s, separates the points of $\mathfrak{M}_s$.

Proposition 6.10. We have

$$\mathfrak{M}_s = \{\varphi \in \mathfrak{M}: |\varphi(s)| = 1, s \in S\}$$

and

$$\Gamma_s = \Gamma.$$

Proof. Since any $s \in S$ is invertible in A_s and $|s| = 1$ on Γ then, according to Proposition 6.7 and Corollary 6.9, for any $\varphi \in \mathfrak{M}_s$, $|\varphi(s)| = 1$, i.e. $\mathfrak{M}_s \subset \{\varphi: |\varphi(s)| = 1\}$.

Let $\varphi \in \mathfrak{M}$ with $|\varphi(s)| = 1$ for any $s \in S$. We define the functional φ_1 on A_s, by

$$\varphi_1(\bar{s}f) = \varphi(\bar{s})\varphi(f) \qquad (s \in S,\ f \in A)$$

and extending continuously to A_s, which is possible as $|\varphi_1(\bar{s}f)| \leqslant \|\bar{s}f\|$.

If $\bar{s}_1 f_1 = \bar{s}_2 f_2$ on Γ, then $s_2 f_1 = s_1 f_2$ on Γ, hence also on $\mathfrak{M}$. Therefore $\varphi(s_2)\varphi(f_1) = \varphi(s_1)\varphi(f_2)$ and, since $|\varphi(s)| = 1$ for any $s \in S$, there results $\overline{\varphi(s_1)}\varphi(f_1) = \overline{\varphi(s_2)}\varphi(f_2)$ which shows that φ_1 is a well defined functional.

We also have

$$\varphi_1(\bar{s}_1 f_1 + \bar{s}_2 f_2) = \varphi_1(\bar{s}_1\bar{s}_2[s_2 f_1 + s_1 f_2]) =$$

$$= \overline{\varphi(s_1 s_2)}\,[\varphi(s_2)\varphi(f_1) + \varphi(s_1)\varphi(f_2)] = \overline{\varphi(s_1)}\varphi(f_1) + \overline{\varphi(s_2)}\varphi(f_2) =$$

$$= \varphi_1(\bar{s}_1 f_1) + \varphi_1(\bar{s}_2 f_2),$$

i.e. φ_1 is linear. At the same time φ_1 is obviously multiplicative and $\varphi_1(1) = 1$, therefore $\varphi_1 \in \mathfrak{M}_s$.

On the other hand for any $f \in A$ we have $\varphi_1(f) = \varphi(f)$ and, since A separates the points of $\mathfrak{M}$, we get $\varphi = \varphi_1 \in \mathfrak{M}_s$.

But Γ_s is the Shilov boundary of A_s, therefore it is determined for A and, since $\Gamma_s \subset \Gamma$, there results $\Gamma_s = \Gamma$.

The proof is complete. ◇

Proposition 6.11. *If S separates the points of Γ, then $A_s = C(\Gamma)$. Therefore we have*

$$\Gamma = \Gamma_s = \mathfrak{M}_s = \{\varphi \in \mathfrak{M} : |(s)| = 1, s \in S\}.$$

Proof. Since S is multiplicatively closed and separates the points of Γ, the set of all elements of the form $g = \bar{s}\Sigma\, c_i s_i$, c_i constants and s, $s_i \in S$, is a subalgebra of $C(\Gamma)$ which separates the points of Γ. This algebra is symmetric, as $\bar{g} = \bar{\sigma}\Sigma \bar{c}_i \sigma_i$ with $\sigma = s_1 \ldots s_n \in S$, $\sigma_i = \sigma s \bar{s}_i \in S$. Then, according to the Stone-Weierstrass theorem, there results $A_s = C(\Gamma)$. ◇

Corollary 6.12. *If S separates Γ then for any $g \in C_R(\Gamma)$, $g \geqslant 0$ and $\varepsilon > 0$, there exists $f \in A$ such that $|g - |f|| < \varepsilon$.*

Proof. As $A_s = C(\Gamma)$, there exist $s \in S$ and $f \in A$ with $|g - \bar{s}f| < \varepsilon$. Then $|g - |f|| = |g - |\bar{s}f|| \leqslant |g - sf| < \varepsilon$. ◇

Proposition 6.13. *If S separates the points of Γ then for any closed subset F of Γ, with $A|F = C(F)$, and any $\mu \in A^{\perp}$ we have $\mu_F = 0$.*

Proof. Let $\mu \in A^{\perp}$, $\|\mu\| = 1$. According to Corollary 4.3, there exists $k < 1$ such that $\|\mu_F\| \leqslant k \|\mu\|$. If we denote by F' the complement of F then, clearly, there is a c such that $\|\mu_F\| \leqslant c \|\mu_{F'}\|$. Let $\varepsilon > 0$ be sufficiently small and a compact $K \subset F'$ such that $\|\mu'_{F'-K}\| < \varepsilon$. Let $g \in C(\Gamma)$, $g \geqslant 0$, with $g = \varepsilon$ on K, $g = 1 - \varepsilon$ on F and $g \leqslant 1 - \varepsilon$ on Γ. Following Corollary 6.12 there exists $f \in A$ such that $g - \varepsilon < |f| < g + \varepsilon$. Hence, we have $\|f\| \leqslant 1$, $|f| \geqslant 1 - 2\varepsilon$ on F and $|f| < 2\varepsilon$ on K.

Since $f\mu \in A^{\perp}$, we obtain

$$(1 - 2\varepsilon) \|\mu_F\| \leqslant \|(f\mu)_F\| \leqslant c\|f\mu)_{F'}\| =$$

$$= c\|f\mu_{F'-K}\| + c\|f\mu_K\| \leqslant c\varepsilon + 2c\varepsilon = 3c\varepsilon.$$

But ε is arbitrary small, therefore $\|\mu_F\| = 0$, i.e. $\mu_F = 0$. ◇

Corollary 6.14. *If S separates the points of Γ and F is a closed subset of Γ for which $A|F = C(F)$, then F is an intersection of peak sets.*

Proof. The proof follows from Proposition 6.13 and Theorem 4.12. ◇

Corollary 6.15. *If S separates the points of Γ then any point of Γ is an intersection of peak sets.*

Therefore the Choquet boundary of A is identical with Γ. Now, let A be a function algebra on X, $\mathfrak{M}$ its maximal ideal space, Γ its Shilov boundary and Σ its Choquet boundary.

A function $s \in A$ will be called (X)-inner if $|s| = 1$ on X.

Let S be a closed multiplicative system of inner function. We assume $1 \in S$. The algebra A is called *S-generated* if the closed subalgebra generated by S in A is identical with A. Since S is multiplicative closed, A is S-generated if and only if the closed linear subspace generated by S in A is identical with A.

As S separates the points of X, from Proposition 6.11 and Corollary 6.15, there results

Proposition 6.16. *Let A be an S-generated algebra on X, $\mathfrak{M}$ its maximal ideal space, Γ its Shilov boundary and Σ its Choquet boundary. We have*

$$\Sigma = \Gamma = X = \{\varphi \in \mathfrak{M}: |\varphi(s)| = 1,\ s \in S\}.$$

We now give an example of an S-generated algebra.

Let G be an abelian group and X the dual of the discrete group G. Let S be a subsemigroup of G such that $G = SS^{-1}$ and $S \cap S^{-1} = \{1\}$. If we consider G as the character group of X then $G \subset C(X)$. Let μ be the normalized Haar measure on X. We write

$$A(S) = \{f \in C(X): \int f\bar{g}\mathrm{d}\mu = 0, g \in G, g \notin S\}.$$

One easily verifies that $A(S)$ is a uniformly closed algebra. As $S \subset A(S)$ and $G = SS^{-1}$, S separates the points of X, which is also true for $A(S)$. Thus, $A(S)$ is a function algebra on X. S is, clearly, a system of inner functions in $A(G)$. We now show that $A(S)$ is S generated.

Let $f \in A(S)$, $f \neq 0$. It is known that for any $\varepsilon > 0$ there exists $u \in C(X)$ such that $\|f*u - f\| < \varepsilon$, where $f*u$ denotes the convolution of f with u. Since the trigonometric polynomials are uniformly dense in $C(X)$, there exists a trigonometric polynomial $v = \Sigma c_i g_i$ such that $\|u - v\| < \varepsilon$. Since

$$f*v = \sum c_i(\int fg_i\mathrm{d}\mu)g_i = \sum_{g_k \in S} c_k g_k$$

and $\|f*v - f\| < 2\varepsilon$, the linear subspace generated by S in $A(S)$, is dense in $A(S)$.

Since $S \cap S^{-1} = \{1\}$ and

$$\int s\mathrm{d}\mu = 0 \qquad (s \in S, s \neq 1),$$

the measure μ is multiplicative on $A(S)$, hence it represents a point φ_0 in the maximal ideal space of $A(S)$.

Theorem 6.17. *The following assertions are equivalent:*

(a) $G = S \cup S^{-1}$,

(b) *$A(S)$ is a Dirichlet algebra on X,*

(c) *$A(S)$ is a logmodular algebra on X.*

(d) *for any $\varphi \in \mathfrak{M}$ there exists a unique representing measure for φ with support in χ,*

(e) *μ is the only representing measure for φ_0 (with support in X),*

(f) *if μ_1 is a representing measure for φ_0, absolutely continuous with respect to μ, then $\mu_1 = \mu$.*

Proof. If $G = S \cup S^{-1}$ then $A(S) + \overline{A(S)}$ contains all trigonometric polynomials, hence it is dense in $C(X)$. Thus, $A(S)$ is clearly a Dirichlet algebra on X and therefore (a) → (b).

The implications (b) → (c) → (d) → (e) → (f) are already known; (f) → (a) follows immediately from Corollary 5.19.

Consider again an S-generated algebra A on X. The system S will be called *analytically free* if for any multiplicative map α from S into the complex plane, with $|\alpha(s)| = 1$, $s \in S$, for any linear combination $\Sigma c_i s_i$ of elements in S, we have

$$|\sum c_i \alpha(s_i)| \leqslant \sum c_i s_i\|.$$

Let G the abelian group $G = S\bar{S}$ and $\hat{G}$ the dual of the discrete group G.

Theorem 6.18. *Let A be an S-generated algebra X with S analytically free. The space X is homeomorphic to the dual $\hat{G}$ of the discrete group G and A is isometrically isomorphic to $A(S)$.*

Proof. For $x \in X$ let α_x be the character on G, defined by

$$\alpha_x(s_1 \bar{s}_2) = s_1(x)\, \overline{s_2(x)}.$$

Obviously $\alpha_x \in \hat{G}$. If $\alpha_{x_1} = \alpha_{x_2}$, then $s(x_1) = s(x_2)$ for any $s \in S$ and, since S separates the points of X, there results $x_1 = x_2$. Therefore the map $x \to \alpha_x$ is an injection of X into $\hat{G}$.

Let us now prove that this map applies X onto $\hat{G}$. If $\alpha \in \hat{G}$ we put

$$\varphi_\alpha(\sum c_i s_i) = \sum c_i \alpha(s_i).$$

But α is a character of the group G, therefore α is a multiplicative map onto S and $|\alpha(s)| = 1$ for any $s \in S$. Since S is analytic free, there results

$$|\sum c_i \alpha(s_i)| \leqslant \|\sum c_i s_i\|$$

hence φ_α can be extended to a multiplicative linear functional on A, i.e. $\varphi_\alpha \in \mathfrak{M}$. Since for any $s \in S$ we have

$$\varphi_\alpha(s) = |\alpha(s)| = 1,$$

from Proposition 6.15 we get $\varphi_\alpha = x \in X$. But, obviously $\alpha_x = x$, therefore the map $x \to \alpha_x$ applies X on $\hat{G}$.

According to the defined topologies on $\mathfrak{M}$ and $\hat{G}$, the map $x \to \alpha_x$ is continuous, hence it is an isomorphism between X and $\hat{G}$.

This homeomophism maps isometrically isomorphic the algebra A in the algebra of functions on $\hat{G}$, generated by the characters of S which, as we have seen, is exactly $A(S)$.

This completes the proof of the theorem. ◇

In the following we give a sufficient condition for a system S of inner function to be analytic free.

For a finite system $\tau = [s_1, s_2, \ldots, s_n]$ of inner functions, we denote by π_τ the map defined on X with values in the n-dimensional torus $T^n = \prod_1^n \{|z| = 1\}$, by

$$\pi_\tau(x) = (s_1(x), s_2(x), \ldots, s_n(x)).$$

Proposition 6.19. *Let S_0 be a system of inner functions such that for any finite system τ of distinct elements of S_0, the map of X into T^n is surjective. Then the multiplicative closed system S generated by S_0 is analytic free.*

Proof. Let α be a multiplicative map of S in T^1 and $s_1, \ldots, s_n$ a finite system of elements in S. Let $\tau = [\sigma_1, \ldots, \sigma_n]$ be a finite system of distinct elements of S_0 such that

$$s_i = \sigma_1^{p_1^i} \sigma_2^{p_2^i} \ldots \sigma_n^{p_n^i} \qquad (i = 1, 2, \ldots, n).$$

Since the map π_τ is surjective, there exists $x \in X$ such that $\sigma_i(x) = \alpha(\sigma_i)$, $i = 1, 2, \ldots, n$. Hence, for any finite system of constants $c_1, c_2, \ldots, c_n$, we have

$$|\sum c_i \alpha(s_i)| = |\sum c_i \alpha(\sigma_1^{p_1^i} \sigma_2^{p_2^i} \ldots \sigma_n^{p_n^i})| =$$

$$= |\sum c_i \alpha(\sigma_1^{p_1^i}) \alpha(\sigma_2^{p_2^i}) \ldots \alpha(\sigma_n^{p_n^i})| =$$

$$= |\sum c_i \sigma_1^{p_1^i}(x) \sigma_2^{p_2^i}(x) \ldots \sigma_n^{p_n^i}(x)| = |\sum c_i s_i(x)| \leqslant \|\sum c_i s_i\|$$

and, therefore, S is analytic free. ◇

Corollary 6.20. *Let* $X = \prod_{1}^{n} \{z\colon |z| = 1\}$, $1 \leqslant n \leqslant \infty$, *and* A *the algebra generated by the coordinate functions* $z_1, z_2, \ldots, z_n$. *Then the multiplicative closed system generated by* $z_1, z_2, \ldots, z_n$ *is analytic free.*

6.3. Maximal algebras

Let A be a function algebra on X. A is said *maximal* (in $C(X)$) if for any function algebra B on X, $A \subset B \subset C(X)$, we have either $A = B$ or $B = C(X)$. The algebra A is called *pervasive* on X if for any closed set $F \subset X$, $F \neq X$, the algebra $A|F$ is dense in $C(F)$. We call A *analytic* on X if any function in A which vanishes on a nonvoid open subset of X is identically zero. We recall here that A is said to be an *antisymmetric* algebra on X if any $f \in A$ with $\bar{f} \in A$ is constant on X. Finally, A is called *essential* on X if the only closed ideal of $C(X)$ in A is the ideal $\{0\}$.

Theorem 6.21. *Let* A *be a function algebra on* X, $A \neq C(X)$. *We have*

a) *If* A *is pervasive then it is analytic.*
b) *If* A *is analytic then it is an integrity domain.*
c) *If* A *is an integrity domain then it is antisymmetric.*
d) *If* A *is antisymmetric then it is essential.*
e) *If* A *is essential and maximal then it is pervasive.*

Proof. a) Let $f \in A$, which vanishes on the non-void open subset U of X and let $K = X - U$. If $g \in C(X)$, since A is pervasive, there exists a sequence $g_n \in A$ such that g_n converges uniformly to g on K. Then, clearly, fg_n converges uniformly to fg on all X. Therefore A contains the (closed) ideal generated by f in $C(X)$.

Let $F = \{x \in X\colon f(x) = 0\}$. Suppose $F \neq X$. It is known that the ideal generated by f in $C(X)$ is equal to the set of all functions in $C(X)$ which vanish on F. Let $g \in C(X)$ and f_n a sequence of functions in A which converges uniformly to g on F. Using eventually a subsequence, we can find a sequence h_n of functions in $C(X)$, converging uniformly on X to the function h and verifying $h_n = g_n$ on F. Then $g_n - h_n$ vanishes on F, hence it belongs to the ideal generated by f in $C(X)$, which is included in A. There results $g_n - h_n \in A$ and, as $g_n \in A$, we get $h_n \in A$ and therefore $h \in A$. But $h = g$ on F, hence $h - g \in A$, that is $g \in A$.

Since g was arbitrarily chosen in $C(X)$, there results $A = C(X)$. This contradicts the hypothesis, therefore $F = X$ and f is identically zero.

b) We assume $fg = 0, f \in A, g \in A$ and let $K = \{x \in X : f(x) = 0\}$. If $U = X - K$ is non-void then $g = 0$ on U and since A is analytic, there follows $g = 0$ on X.

c) Let $B = \{f \in A : \bar{f} \in A\}$. We define on X the equivalence relation $x \sim y$ if $f(x) = f(y)$ for any $f \in B$.

Let Y be the compact space obtained by the factorisation of X by this equivalence relation.

Consider $\hat{B} = \{\hat{f} \in C(Y) : \hat{f}(\hat{x}) = f(x), f \in B\}$. One can easily verify that $\hat{B}$ is a symmetric function algebra on Y. From the Stone-Weiersrass theorem there follows $\hat{B} = C(Y)$. If Y does not reduce to one point then there are $\hat{f}_1, \hat{f}_2 \in \hat{B}$ such that $\hat{f}_1 \neq 0$, $\hat{f}_2 \neq 0$ and $\hat{f}_1 \hat{f}_2 = 0$. Then, clearly, $f_1 \neq 0$, $f_2 \neq 0$ and $f_1 f_2 = 0$, where f_1, f_2 are those functions in B which satisfy $\hat{f}_1(\hat{x}) = f_1(x)$, $\hat{f}_2(\hat{x}) = f_2(x)$. But this contradicts the fact that A is an integrity domain. Therefore V reduces to one point, hence B reduces to constant functions.

d) Let us assume that A contains a closed ideal I of $C(X)$. These ideals, as it is known, are of the form

$$I = \{f \in C(X) : f = 0 \text{ on } F\}$$

with F a closed subset of X. If I does not reduce to 0, then $F \neq X$. Hence there exists a real continuous function, which is not identically zero but vanishes on F. Yet this function belongs to I, therefore to A, which contradicts the antisymmetry of A.

e) Let $F \subset X$, $F \neq X$, F closed. Since A is essential there exists $g \in C(X)$ with $g = 0$ on F and $g \notin A$. Since A is maximal, the closed algebra generated by A and g is identical with $C(X)$.

Consider $h \in C(X)$ of the form

$$h = \lim \left(f_0^{(n)} + f_1^{(n)} g + \ldots + f_k^{(n)} g^{k_n}\right),$$

where $f_i^{(n)} \in A$. Then, clearly, we have

$$h \mid F = \lim f_0^{(n)} \mid F$$

and therefore A is pervasive.

The theorem is proved. ◇

Theorem 6.22. *Let A be a pervasive algebra on X, $\mathfrak{M}$ its maximal ideal space, Γ its Shilov boundary and Σ its Choquet boundary. For any $\varphi \in \mathfrak{M}$ there exists a unique representing measure with support in X. Therefore $\Sigma = \Gamma = X$. If $\varphi \in \mathfrak{M} - X$ then the representing measure of φ has its support identical with X.*

Proof. Two representing measure μ and μ' for the same $\varphi \in \mathfrak{M}$ satisfy $\mu_F = \mu'_F$ for any closed subset $F \subset X$, $F \neq X$. As $\mu(X) = \mu'(X) = 1$, there result $\mu = \mu'$.

Now let $\varphi \in \mathfrak{M} - X$, μ its representing measure and F its support. If $F \neq X$, since $\mu_F = \mu$, μ results multiplicative on $C(F)$; hence there exists $x \in F$ such that

$$\varphi(f) = \int f \mathrm{d}\mu = f(x) \qquad (f \in A)$$

which is impossible as A separates $\mathfrak{M}$ and $\varphi \in \mathfrak{M} - X$. ◇

Theorem 6.23. *Let A be pervasive on X and $\varphi \in A$. If there exists $\varphi_0 \in \mathfrak{M} - X$ such that*

$$|f(\varphi_0)| = \|f\|,$$

then f is constant on M.

Proof. Let $f(\varphi_0) = e^{i\alpha}\|f\|$ and $g = e^{-i\alpha}f$, $g = u + iv$. Since

$$g(\varphi_0) = u(\varphi_0) + iv(\varphi_0) = \|f\|$$

there results $u(\varphi_0) = \|f\|$. Since $\|g\| = \|f\|$ there follows $u(\varphi) \leqslant u(\varphi_0)$ for any $\varphi \in \mathfrak{M}$.

Let $U = \{x \in X : u(x) < u(\varphi_0)\}$ and μ be the representing measure of φ_0. We then have

$$u(\varphi_0) = \int u \mathrm{d}\mu = \int_U u \mathrm{d}\mu + \int_{X-U} u \mathrm{d}\mu < u(\varphi_0)\mu(\mathrm{U}) + u(\varphi_0)\mu(X - U) = u(\varphi_0)$$

which is impossible. We used the fact that X is the support of μ. Therefore u is constant and, since A pervasive is also antisymmetric, g is constant, hence f is constant. ◇

6.4. Function algebras on compact sets of the complex plane

Let X be a compact set of the complex plane. We denote by $P(X)$ the subalgebra of $C(X)$ obtained by closing in $C(X)$ the set of polynomials on X. By $R(X)$ we denote the algebra of all functions in $C(X)$ which are uniform limits on X of rational functions with poles outside X; $A(X)$ will denote the algebra of all functions in $C(X)$, analytic in any interior point of X.

One easily verifies that $P(X)$, $R(X)$, $A(X)$ are function algebras on X and $P(X) \subset R(X) \subset A(X)$.

Proposition 6.24. *The maximal ideal space of* $P(X)$ *is homeomorphic to the polynomially convex hull* $\hat{X}$ *of* X.

Proof. It is known that

$$\hat{X} = \{z \in C\colon |P(z)| \leqslant \sup_{x \in X} |P(x)| \text{ for any polynomial } P\}.$$

Then X is a determining set for $P(\hat{X})$ and, since $P(X) = P(\hat{X}) \mid X$, from Theorem 2.17 it follows that $P(\hat{X})$ and $P(X)$ are isomorphic, hence their maximal ideal spaces are homeomorphic. Let $\mathfrak{M}$ be the maximal ideal space of $P(\hat{X})$, which, by the natural homeomorphism, is equal to the maximal ideal space of $P(X)$. We know that $\hat{X}$ is homeomorphically embedded in $\mathfrak{M}$ by means of the mapping $x \to e_x$, where

$$e_x(f) = f(x) \qquad (f \in P(\hat{X})).$$

We now show that the map $x \to e_x$ applies $\hat{X}$ on $\mathfrak{M}$. Let $\varphi \in \mathfrak{M}$ and let z be the identity map of the complex plane, which obviously belongs to $P(\hat{X})$. Let $x_0 = \varphi(z)$.

Since for any polynomial P we have $\varphi(P) = P(\varphi(z))$ there results

$$|P(x_0)| = |P(\varphi(z))| = |\varphi(P)| \leqslant \sup_{x \in X} |P(x)|.$$

Therefore $x_0 \in \hat{X}$ and

$$f(x_0) = \varphi(f) \qquad (f \in P(\hat{X}))$$

that is $e_{x_0} = \varphi$. Then $x \to e_x$ applies $\hat{X}$ on $\mathfrak{M}$.

The proof is complete. ◇

We notice that, according to the maximum modulus principle $\hat{X}$ is the union of X with the bounded connected components of the complement of X.

Let us suppose that X belongs to the boundary of the unbounded connected component of its complement. In this case we shall prove that $P(X)$ is a Dirichlet algebra on X.

We first prove two helping lemmas (see CARLESON, [2]). Let $\mathrm{d}\lambda$ be the Lebesgue measure in the plane.

Lemma 6.25. *Let X be a compact set of the complex plane and Ω the unbounded connected component of the complement of X. Let μ be a real measure on X. Then*

$$u(z) = \int_X \log \frac{1}{|z-x|} \, \mathrm{d}\mu(x)$$

converges absolutely almost everywhere with respect to the Lebesgue measure $\mathrm{d}\lambda$. If $u(z) = 0$ in Ω, then $u(z) = 0$ in any point $z \in \bar{\Omega}$ in which we have absolute convergence.

Proof. Since the function $\log \frac{1}{|z|}$ is locally integrable with respect to $\mathrm{d}\lambda$ and the measure $\mathrm{d}\,|\mu|$ is finite, there results

$$\int_{|z|<R} \left\{ \int_X \log \frac{1}{|z-x|} \, \mathrm{d}|\mu|(x) \right\} \mathrm{d}\lambda(z) < \infty$$

and therefore

$$\int_X \log \frac{1}{|z-x|} \, \mathrm{d}|\mu|(x) < \infty,$$

λ-almost everywhere. Hence, the absolute convergence λ-almost everywhere of $u(z)$ follows.

Now assume $u(z) = 0$ in Ω. Let $z_0 \in \partial\Omega$ (the topological boundary of Ω) in which $u(z)$ converges absolutely.

To simplify the notation, we assume $z_0 = 0$, which does not affect the generality of the proof.

Let $\delta > 0$ and $0 < r_1 < r_2 \leqslant \delta$. As $0 \in \partial\Omega$ we can find a positive measure σ with support in Ω such that

$$\sigma(\{z : r_1 < |z| < r_2\}) = r_2 - r_1$$

and $\sigma(K) = 0$ for any compact K situated outside the circle $|z| \leqslant \delta$.

Since $u(z) = 0$ in G we have

$$\frac{1}{\delta} \int u(z)\, d\sigma(z) = 0.$$

We fix a $\rho > 0$. Then

$$\int\limits_{|x|<\rho} \left\{ \frac{1}{\delta} \int \log \frac{1}{|z-x|}\, d\sigma(z) \right\} d\mu(x) +$$

$$+ \int\limits_{|x|<\rho} \left\{ \frac{1}{\delta} \int \log \frac{1}{|z-x|}\, d\sigma(z) \right\} d\mu(x) = \frac{1}{\delta} \int u(z)\, d\sigma(z) = 0.$$

By the construction of the measure $d\sigma$, there results that for $|x| \geqslant \rho$ we have

$$\lim_{\delta \to 0} \frac{1}{\delta} \int \log \frac{1}{|z-x|}\, d\sigma(z) = \log \frac{1}{|x|} \cdot$$

If $|x| < \rho$ we have:

$$\frac{1}{\delta} \int \log \frac{1}{|z-x|}\, d\sigma(z) \leqslant \frac{1}{\delta} \int \log \frac{1}{\big||z|-|x|\big|}\, d\sigma(z) =$$

$$= \frac{1}{\delta} \int_0^\delta \log \frac{1}{r-|x|}\, dr = \log \frac{1}{|x|} + \int_0^\delta \log \frac{1}{\left|1 - \dfrac{2}{|z|}\right|}\, dr \leqslant \log \frac{1}{|x|} + C,$$

where

$$C = \sup_{T>0} \frac{1}{T} \int_0^T \log \frac{1}{|1-t|} \, dt < \infty.$$

Hence

$$\int_{|x|>\rho} \left\{ \frac{1}{\delta} \int \log \frac{1}{|z-x|} \, d\sigma(z) \right\} d\mu(x) \leqslant \left| \int \left(\log \frac{1}{|x|} + C \right) d|\mu|(x) \right|$$

and therefore, since $\delta \to 0$, we obtain

$$\int_{x \geqslant \rho} \log \frac{1}{|x|} \, d\mu(x) \leqslant \int_{x \geqslant \rho} \left(\log \frac{1}{|x|} + C \right) d|\mu|\,(x).$$

Since 0 is assumed to be an absolute convergence point for $u(z)$ the inequality's right handside tends to 0 when $\rho \to 0$, hence

$$u(0) = \lim_{\rho \to 0} \int_{|x| \geqslant \rho} \log \frac{1}{|x|} \, d\mu(x) = 0.$$

This completes the proof of Lemma 6.25. ◇

Lemma 6.26. *Let X, μ and u be as in the preceding lemma. If $u(z) = 0$ λ-almost everywhere in all the plane, then $\mu = 0$.*

Proof. For any function g of class C^2 and compact support we know that (see Stoilow, S., [1]):

$$g(x) = -\frac{1}{2\pi} \int \Delta g(z) \log \frac{1}{|z-x|} \, d\lambda(z).$$

By Fubini theorem we get

$$\int_X g(x) d\mu(x) = \frac{1}{2\pi} \int \Delta g(z) \left\{ \int_X \log \frac{1}{|z-x|} \, d\mu(x) \right\} d\lambda(z) =$$

$$= -\frac{1}{2\pi} \int g(z) u(z) d\lambda(z) = 0.$$

Since any function in $C_R(X)$ is a uniform limit on X of functions with compact support and of class C^2, there results $\mu = 0$. ◇

Theorem 6.27. *Let X be a compact set in the plane, contained in the boundary of the unbounded connected component Ω of its complement. The algebra $P(X)$ is a Dirichlet algebra on X.*

Proof. Let μ be a real measure on X, orthogonal on $\mathrm{Re}P(X)$.

The logarithmic potential

$$u(z) = \int_X \log \frac{1}{|z-x|}\, \mathrm{d}\mu(x)$$

is known as an harmonic function outside X (see BRELOT, M., [1]). For any z with the property $|z| > \sup_{x \in X} |x|$, the function

$$f(x) = \log \frac{1}{z-x}$$

belongs to $P(X)$, since the series $\sum_{n=1}^{\infty} \frac{(-1)^n}{n} \left(\frac{x}{z}\right)^n$ converges uniformly on X. As $\log \frac{1}{|z-x|} = \mathrm{Re} f(x)$, we get $u(z) = 0$ for any z with $|z| > \sup_{x \in X} |x|$. But since $u(z)$ is harmonic in Ω and $\{z\colon |z| > \sup_{x \in X} |x|\} \subset \Omega$ there follows $u(z) = 0$ for any $z \in \bar{\Omega}$. From Lemma 6.25 there results $u(z) = 0$ in any point of Ω in which $u(z)$ converges absolutely.

Let D be a bounded connected component of the complement of X and $a \in D$. Let $Y = D \cup X$. Since D is bounded and $\partial D \subset X$, Y is a compact set in the complex plane. We easily see that Ω is the unbounded connected component of the complement of Y.

Let $A(Y)$ be the algebra of continuous functions on Y, which are analytic in any interior point of Y. According to the maximum modulus theorem X is a determining set for $A(Y)$. Hence there exists a positive measure λ_a on Y such that

$$f(a) = \int_X f \mathrm{d}\lambda_a \qquad (f \in A(Y)).$$

Let $\nu_a = \varepsilon_a - \lambda_a \cdot \nu_a$ is a real measure on Y.

If $z \in \Omega$, then the function $f(y) = \log \dfrac{1}{z - y}$ obviously belongs to $A(Y)$, therefore

$$u_a(z) = \int_X \log \frac{1}{|z - y|} \, \mathrm{d}v_a(y) = \log \frac{1}{|z - a|} - \int_X \log \frac{1}{|z - x|} \, \mathrm{d}\lambda_a(x) = 0.$$

If $z_0 \in \partial\Omega$, since the potential is lower semicontinuous, there follows

$$|u_a(z_0)| \leqslant \int \log \frac{1}{|z_0 - y|} \, \mathrm{d}|y_a| \leqslant \left| \log \frac{1}{|z - a|} \right| +$$

$$+ \lim_{\substack{z \to z_0 \\ z \in \Omega}} \int_X \log \frac{1}{|z - x|} \, \mathrm{d}\lambda(x) \leqslant 2 \left| \log \frac{1}{|z_0 - a|} \right|$$

hence $u(z_0)$ converges absolutely in any point z_0 of $\partial\Omega$. From Lemma 6.25 there results $u_a(z) = 0$ on $\partial\Omega$.

Thus, for any $z \in X$ we have

$$\log \frac{1}{|z - a|} = \int \log \frac{1}{|z - x|} \, \mathrm{d}\lambda_a(x)$$

and then

$$\int_X \left\{ \int_X \log \frac{1}{|z - x|} \, \mathrm{d}|\mu|(x) \right\} \mathrm{d}\lambda_a(z) =$$

$$= \int_X \left\{ \int_X \log \frac{1}{|z - x|} \, \mathrm{d}\lambda_a(z) \right\} \mathrm{d}|\mu|(x) =$$

$$= \log \int_X \log \frac{1}{|z - a|} \, \mathrm{d}|\mu|(x) < \infty.$$

Therefore the set of points in X in which $u(z)$ is divergent has the measure zero with respect to λ_a.

Since $u(z) = 0$ in any point $z \in X$ in which $u(z)$ converges absolutely, we obtain

$$0 = \int_X u(z) d\lambda_a(z) = \int_X \left\{ \int_X \log \frac{1}{|z-x|} \, d\mu(x) \right\} d\lambda_a(z) =$$

$$= \int_X \left\{ \int_X \log \frac{1}{|z-x|} \, d\lambda_a(z) \right\} d\mu(x) = \int \log \frac{1}{|z-a|} d\mu(x) = u(a).$$

Hence u vanishes in any point belonging to the bounded connected components of the complement of X. Since $u(z) = 0$ λ-almost everywhere on $\bar{\Omega}$, $u(z) = 0$ λ-almost everywhere on the entire plane and therefore, using lemma 6.26, $\mu = 0$. ◇

Proposition 6.28. *The maximal ideal space of $R(X)$ is homeomorphic to X.*

Proof. Let $\mathfrak{M}$ be the maximal ideal space of $R(X)$. We know that X is homeomorphically embedded in $\mathfrak{M}$ by means of the map $x \to e_x$ where e_x is defined by

$$e_x(f) = f(x) \qquad (f \in R(X)).$$

We shall prove that this map applied X on $\mathfrak{M}$. Let z be the identity map of the complex plane. Obviously $z \in R(X)$. Let $\varphi \in \mathfrak{M}$ and $\alpha = \varphi(z)$. If $\alpha \notin X$ then $(z - \alpha)^{-1} \in R(X)$, hence

$$1 = \varphi(1) = \varphi[(z - \alpha)(z - \alpha)^{-1}] = (\varphi(z) - \alpha)\varphi(z - \alpha)^{-1} = 0$$

which is impossible. Therefore $\alpha \in X$. It is then clear that

$$e_\alpha(f) = f(\alpha) = f(\varphi(z)) = \varphi(f)$$

for any rational function of $R(X)$, hence for any function of $R(X)$. There results $\varphi = e_a$ and therefore the map $x \to e_x$ is surjective.

Lemma 6.29. *Let μ be a measure on X. Let*

$$N(y) = \int_X \frac{d|\mu|}{|x-y|}.$$

Then $N(y)$ is finite almost everywhere with respect to λ. If

$$F(y) = \int_X \frac{d\mu(x)}{x - y}$$

vanishes almost everywhere with respect to λ, then $\mu = 0$. If $\mu \in R(X)^{\perp}$, then $U = \{y \in C: N(y) < \infty,\ F(y) \neq 0\} \subset X$ and for any $y \in U$, the measure μ_y defined by $d\mu_y = \frac{1}{F(y)}(z-y)^{-1}d\mu(z)$, verifies the relation

$$f(y) = \int f d\mu_y \qquad (f \in R(X)).$$

Proof. The first part of the lemma follows from the fact that $\frac{1}{|z|}$ is an integrable function with respect to λ on any compact in plane, and $|\mu|$ is a finite measure on X.

Assume now $F(y) = 0$ λ-almost everywhere.

Let g be a continuous function with compact support and indefinitely differentiable. It is then known that

$$g(y) = -\frac{1}{2\pi}\int (y-w)^{-1}\left(\frac{\partial}{\partial x} + i\frac{\partial}{\partial y}\right) g(w) d\lambda(w).$$

By integrating relative to $d\mu$ and applying Fubini's theorem we get

$$\int_X g(y)d\mu(y) = -\frac{1}{\pi}\int_X \left\{\int (y-w)^{-1}\frac{1}{2}\left(\frac{\partial}{\partial x} + i\frac{\partial}{\partial y}\right) g(w)d\lambda(w)\right\} \times$$

$$\times\, d\mu(y) = -\frac{1}{2\pi}\int \left(\frac{\partial}{\partial x} + i\frac{\partial}{\partial y}\right) g(w) \int (y-w)^{-1} d\mu(y) d\lambda(w) =$$

$$= -\frac{1}{2\pi} F(y) \int \left(\frac{\partial}{\partial x} + i\frac{\partial}{\partial y}\right) g(w) d\lambda(w) = 0.$$

Since any continuous function on X may be uniformly approximated on X with continuous functions with compact support and indefinitely differentiable, $\mu = 0$.

Suppose now $\mu \in R(X)^{\perp}$ and let $y \in U$. If $y \notin X$ then $f(x) = (x-y)^{-1} \in R(X)$, hence

$$F(y) = \int_X (x-y)^{-1} d\mu = 0$$

which is impossible. Therefore $U \subset X$.

Let $y \in U$. Since $F(y) \neq 0$ and $(z-y)^{-1}$ is integrable with respect to μ on X, then μ_y defined by

$$d\mu_y = \frac{1}{F(y)} (z-y)^{-1} d\mu$$

is a measure on X. Since for any rational function of $R(X)$, the function

$$h(x) = \frac{f(x) - f(y)}{x - y}$$

is also a rational function in $R(X)$, there results

$$f(y) - \int_X f d\mu_y = f(y) - \frac{1}{F(y)} \int f(x)\,(x-y)^{-1} d\mu(x) =$$

$$= -\frac{1}{F(y)} \left[\int \frac{f(x) - f(y)}{x - y} d\mu(x) \right] = 0$$

for any rational function in $R(X)$ and, by continuity, we get

$$f(y) = \int f d\mu_y, \qquad (f \in R(X)).$$

The lemma is completely proved. ◇

Let Σ be the Choquet boundary of $R(X)$. Since X is metrisable, any point of Σ is a peak point for $R(X)$.

Proposition 6.30. *Let $x \in X - \Sigma$ and let Δ be the Gleason part of X which contains x. Then $\lambda(\Delta) > 0$.*

Proof. According to Theorem 2.23 there exists $\mu \ll \varepsilon_x$ such that $\mu(\{x\}) = 0$. Let $\nu = \mu - \varepsilon_x$. Then clearly $\nu \in R(X)^{\perp}$, $\nu \neq 0$. Let F

and U be as in Lemma 6.29. From Lemma 6.29, there results $U \subset X$ and that F does not vanish everywhere with respect to λ. Then

$$\lambda(U) = \lambda\{y \in C: F(y) \neq 0\} > 0.$$

We now show that $U \subset \Delta$. Let $y \in U$.

According to Lemma 6.29, the measure ν_y defined by

$$d\nu_y = \frac{1}{F(y)}(z-y)^{-1} d\nu$$

satisfies the relation

$$f(y) = \int f d\nu_y \qquad (f \in R(X)).$$

On the other hand, from Theorem 3.15 there follows a representing measure for y, absolutely continuous with respect to ν_y, hence absolutely continuous with respect to ν. This measure must obviously be of the form $h d\mu + c\varepsilon_y$ with $h \in L^1(d\mu)$. Then $\mu_y = (h+c)\mu$ is a representing measure for y, absolutely continuous with respect to μ. Then, from Proposition 6.5 we get $y \in \Delta$.

Hence $U \subset \Delta$ and therefore $\lambda(\Delta) \geqslant \lambda(U) > 0$. ◇

Corollary 6.31. *There exists at most a countable set of Gleason parts of X (with respect to $R(X)$), which do not reduce to one point.*

Proof. Since the points of Σ form point Gleason parts (Corollary 3.14), the proof follows immediately. ◇

Corollary 6.32. *A point $x \in X$ forms a point Gleason part of X with respect to $R(X)$ if and only if it is a peak point for $R(X)$. Hence the set P of all points $x \in X$ which form point Gleason parts is identical to Σ.*

Theorem 6.33. *Let X be a compact set of the complex plane, Σ its Choquet boundary with respect to $R(X)$ and P the set of all the points of X which form point Gleason parts of X (with respect to $R(X)$). The following assertions are equivalent.*

(a) $R(X) = C(X)$

(b) $X = P$

(c) $X = \Sigma$.

Proof. We already know that (a) → (c) and that (b) and (c) are equivalent. It remains to prove (c) → (a). Let $\mu \in R(X)^{\perp}$. If $\mu \neq 0$ then, according to Lemma 6.29, there exists $y \in X$ such that $F(y) \neq 0$ with F constructed as in the mentioned lemma, and the measure μ_y defined by

$$\mathrm{d}\mu_y = \frac{1}{F(y)}(z - y)^{-1}\mathrm{d}\mu$$

satisfies the relation

$$f(y) = \int f\mathrm{d}\mu_y \qquad (f \in R(X)).$$

Now, according to Theorem 3.15, there exists a representing measure for y, absolutely continuous with respect to μ_y, hence absolutely continuous with respect to μ. Since $y \in \Sigma$, ε_y is the only representing measure for y, hence ε_y is absolutely continuous with respect to μ, that is $\mu(\{y\}) \neq 0$.

Since y is a peak point for $R(X)$, there exists $f \in R(X)$ such that $f(y) = 1$ and $|f| < 1$ on $X - \{y\}$. It is then clear that for sufficiently large n

$$\int f^n \,\mathrm{d}\mu \neq 0$$

which contradicts the fact that $\mu \in R(X)^{\perp}$. Then $\mu = 0$, i.e. $R(X) = C(X)$.

The theorem is proved. ◇

Let now X be a compact set of the complex plane, contained in the boundary of the unbounded connected component of its complement.

In this case, according to Theorem 6.27, $P(X)$ is a Dirichlet algebra on X.

As $P(X) \subset R(X)$, $R(X)$ is also a Dirichlet algebra on X and, following Corollary 6.2, the Choquet boundary of $R(X)$ is identical to X. Then, from Theorem 6.33 there results:

Corollary 6.34. *If X is contained in the boundary of the umbounded connected component of its complement, then $R(X) = C(X)$.*

Proposition 6.35. *Let X be a compact set of the complex plane, contained in the boundary of the unbounded connected component of* *its complement. If $\mu \in P(X)^{\perp}$ and it is singular with respect to any positive measure on X and multiplicative on P(X) then $\mu = 0$.*

Proof. We have seen that the maximal ideal space of $P(X)$ is the polynomially convex hull $\hat{X}$ of X. Hence, if $z \notin \hat{X}$, then the function $f(x) = \frac{1}{x-z} \in P(X)$, that is

$$\int \frac{1}{(x-z)^n} \, d\mu = 0 \qquad (z \notin \hat{X},\ n = 1, 2, \ldots).$$

Let now $z \in \hat{X}$, $z \notin X$ and μ_z be the representing measure of z with support in X. We put

$$d\nu = \frac{d\mu}{z - x} - c\, d\mu_z$$

where

$$c = \int \frac{d\mu(x)}{x-z}.$$

For $n \geqslant 1$ we have

$$\int (x-z)^n d\nu(x) = \int (x-z)^{n-1} d\mu(x) - c \int (x-z)^n d\mu_z(x) = 0$$

since $\mu \in P(X)^{\perp}$ and μ_z is a representing measure for z. As obviously $\nu(1) = 0$, $\nu \in P(X)^{\perp}$. Let $d\nu = d\nu_a + d\nu_s$ be the Lebesgue decomposition of ν with respect to μ_z. Since $P(X)$ is a Dirichlet algebra on X, μ_z is the only representing measure for z with support in X, hence according to the M. and F. Riesz Theorem (Theorem 5.6), $\nu \in P(X)^{\perp}$. But μ is singular with respect to μ_z, therefore $d\nu_s = (x-z)^{-1} \, d\mu$. We then have

$$\int \frac{d\mu(x)}{x-z} = 0.$$

Now, if we repeat the procedure, taking $(x-z)^{-1} \, d\mu \in P(X)$ instead of μ, we obtain $(x-z)^{-2} d\mu \in P(X)^{\mathrm{T}}$, and so on. Hence

$$\int \frac{d\mu(x)}{(x-z)^n} = 0 \qquad (z \in \hat{X} - X,\ n = 1, 2, \ldots).$$

Then μ results orthogonal on any rational function with poles outside X and, from Corollary 4.33, we get $\mu = 0$.

The proof is therefore complete. ◇

We are now able to give an example of a Dirichlet algebra A with the following properties: there exist measures $\mu \in A^\perp$ which are absolutely continuous with respect to no representing measure of A and, any measure of $A^\perp$, singular with respect to any representing measure of A, is 0.

$$\text{Let } X = \{z: |z| = 1\} \cup \{z: |z-2| = 1\} \text{ and } A = P(X).$$

Since, clearly, X belongs to the boundary of the unbounded connected component of its complement, according to Theorem 6.27, $P(X)$ is a Dirichlet algebra on X and following Proposition 6.35, any measure of $P(X)^\perp$, singular with respect to all representing measures for A, is 0. On the other hand, the complex measure dz is obviously orthogonal on $P(X)$ and, since any representing measure for $P(X)$ has its support either in $\{z: |z| = 1\}$ or in $\{z: |z-2| = 1\}$, dz is absolutely continuous with respect to none of these representing measures.

Theorem 6.36. (Mergelyan). *Let X be a compact set of the complex plane with a connected complement. Then any continuous function on X, analytic in any interior point of X, is a uniform limit, on X, of polynomials, i.e. $P(X) = A(X)$.*

Proof. Let $Y = \partial X$ (the topological boundary of X). Since X has a connected complement, Y belongs to the unbounded connected component of its complement. According to the maximum modulus theorem, Y is a determining set for $P(X)$, hence $P(X)$ is isometrically isomorphic to $P(X) \mid Y = P(Y)$.

At the same time Y is also a determining set for $A(X)$, hence $A(X)$ and $A(X) \mid Y$ are isometrically isomorphic.

Now, let μ be a measure on Y, orthogonal on $P(Y)$. Since $P(Y)$ is a Dirichlet algebra on Y (Theorem 6.27) from Theorem 6.6 there results

$$d\mu = \sum_{i=1}^{\infty} h_i d\lambda_i + \sigma$$

where λ_i are representing measures for $P(Y)$, $h_i \in H^1_{\lambda_i}$ and $\sigma \in P(Y)^\perp$ is singular with respect to any representing measure of $P(Y)$.

From Theorem 6.35 there follows $\sigma = 0$.

Since X has a connected complement, X is polyomially convex, hence (Proposition 6.24) the maximal ideal space of $P(X)$ is homeomorphic to X and also homeomorphic to the maximal ideal space of $P(Y)$.

Assume λ_i represents the point $x_i \in X$. Since Y is determining for $A(X)$ too, there exists a representing measure λ_i' for x_i (with respect to $A(X)$), with support in Y. But $P(X) \subset A(X)$, therefore λ_i' is a representing measure for x_i with respect to $P(X)$, hence also with respect to $P(Y)$; as $P(Y)$ is a Dirichlet algebra on y there results $\lambda_i' = \lambda_i$

Thus, λ_i is a representing measure for x_i with respect to $A(X)$, hence also with respect to $A(X)| Y$. Since $P(X) \subset A(X) | Y$, $A(X)| Y$ is a Dirichlet algebra on Y. It is clear that $H^1_{\lambda_i}$ relative to $P(Y)$ is included in $H^1_{\lambda_i}$ relative to $A(X) | Y$, hence $h_i \in H^1_{\lambda_i}$ relative to $A(X) | Y$. Then for any $f \in A(X) | Y$, $fh_i \in H^1_{\lambda_i}$ (relative to $A(X) | Y$), and therefore

$$\int fh_i \mathrm{d}\lambda_i = 0 \qquad (f \in A(X)| Y).$$

Hence

$$\int f \mathrm{d}\mu = \sum_{i=1}^{\infty} \int fh_i \mathrm{d}\lambda_i = 0 \qquad (f \in A(X)| Y)$$

that is $\mu \in (A(X) | Y)^{\perp}$.

There results $(A(X) | Y) = P(Y)$, hence $A(X) = P(X)$.

The proof is therefore complete. ◇

Corollary 6.37. *If X has a connected complement then the maximal ideal space of $A(X)$ is homeomorphic to X.*

6.5. The standard algebra and H^∞ algebra

Let $D = \{z \in C: |z| < 1\}$ and $T = \partial \bar{D} = \{z \in C: |z| = 1\}$. From the previous paragraph we know that $A(\bar{D}) = P(D)$ is isometrically isomorphic to $A(\bar{D})| T = P(T)$. We denote this algebra by A and consider it is a function algebra on T. As we have already mentioned, A will be called the *standard algebra.*

We have seen that the maximal ideal space of A is homeomorphic to $\bar{D}$. At the same time A is a Dirichlet algebra on T, hence T is identical to Choquet and Shilov boundaries of A.

The multiplicative closed system of inner functions $S = \{1, z, z^2, \ldots\}$ is analytic free and generates A, hence A is identical to $A(S)$. Therefore, a function $f \in C(T)$ belongs to A if and only if

$$\int f z^n d\mu = 0 \qquad (n = 1, 2, \ldots)$$

where μ is the normalized Lebesgue measure on T (which, as is already known, is also the Haar measure of T considered as the dual of the discrete group of the additive group of integers).

Proposition 6.38. *A is a maximal subalgebra of $C(T)$.*

Proof. Let B be a function algebra on T such that $A \subset B$ and μ is the normalized Lebesgue measure on T.

If μ is multiplicative on B, then for any $f \in B$ we have

$$\int f z^n d\mu = \int f d\mu \int z^n d\mu = 0 \qquad (n = 1, 2, \ldots)$$

hence $f \in A$. There results $B = A$.

Let λ be a representing measure for B. If

$$\int z d\lambda = 0,$$

then clearly λ is a representing measure of the point 0 (with respect to A) and, as A is Dirichlet, $\lambda = \mu$.

Thus, if μ is not multiplicative on B, there results

$$\int z d\lambda \neq 0$$

for any representing measure of B, hence z does not vanish, on the maximal ideal space of B. Therefore $\bar{z} = z^{-1}$ belongs to B and from the Stone-Weierstrass theorem we obtain $B = C(T)$.

The proposition is proved. ◇

Since the algebra A is antisymmetric, it is also essential and from maximality, there results it is pervasive (Theorem 6.21).

Therefore A has all the restrictive properties required until now in the frame of this theory to function algebras, hence all the established results hold for A too.

If μ is the normalized Lebesgue measure on X, we denote by H^p the space $H^p(d\mu)$. According to Fatou theorem, H^p is the set of all functions in L^p which may be analytically extended in D. We shall complete the theory of H^p-space given in chapter 5 by the following results valid in the case of the standard algebra.

Proposition 6.39. *Let S be an invariant subspace of L^p. S is simply invariant if and only if $zS \neq S$.*

The proof is immediate.

If an invariant subspace S is not simply invariant, then $zS = S$, hence $\bar{z}S \subset S$.

Proposition 6.40. *The algebra H^∞ is a weakly maximal subalgebra of the algebra L^∞.*

Proof. Let B be a weakly closed subalgebra of the algebra L^∞ such that $H \subset B$. Then B is obviously an invariant subspace of L^∞. If B is not simply invariant then $\bar{z}B \subset B$, hence $\bar{z} \in B$. But B is a weakly closed subalgebra, then $B = L^\infty$. If B is simply invariant then, according to the theorem of invariant subspaces, there exists a function $F \in B$, $|F| = 1$, such that $B = FH^\infty$. Since $F \in B$, there exists $h \in H^\infty$ such that $F^2 = Fh$ and therefore $F = h \in H^\infty$, that is $B = H^\infty$.

In § 5.6 we have seen that H^∞ may be considered as a function algebra on the maximal ideal space of L^∞. From Theorem 5.24, H^∞ is even a logmodular algebra on this space.

In the following, X will denote the maximal ideal space of L^∞. We know that it is a totally nonconnected compact space. From this remark and from the following proposition it results that H^∞ is not a Dirichlet algebra on X.

Proposition 6.41. *If A is a Dirichlet algebra on X and X is totally nonconnected, then $A = C(X)$.*

Proof. Let F be a closed subset of X and $f \in A$ such that $|1 - \text{Re} f| < \frac{1}{2}$ on F and $|\text{Re} f| < \frac{1}{2}$ on $X - F$. Such an f does exist in A as $\chi_F \in C_R(X)$ and A is Dirichlet on X.

Let $Y_1 = f(K)$ and $Y_2 = f(X - K)$. Y_1 and Y_2 are compact sets of the complex plane and we have $|\text{Re} z| < \frac{1}{2}$ for $z \in Y_1$ and $|\text{Re} z| > \frac{1}{2}$

for $z \in Y_2$. Then, there exist two disjoint rectangles R_1 and R_2 such that $Y_1 \subset R_1$, $Y_2 \subset R_2$. Therefore, $R = R_1 \cup R_2$ is a compact set of the plane, with a connected complement, hence there exists a sequence of polynomials p_n which converges uniformly on R to $\chi_{R_1'}$. Then, clearly, $p_n(f)$ converges uniformly on X to χ_F, hence $\chi_F \in A$.

Therefore A contains the function χ_F for any closed subset F of X, hence it is clear that $A = C(X)$.

The proof is complete. ◇

So H^∞ is a logmodular algebra on X which is not a Dirichlet algebra.

Following Proposition 6.16, H^∞ can not be generated by a free analytic system of inner functions. It remains an open problem if H^∞ is generated by the set of its inner functions or if these functions at least separate the points of X.

Notes

Dirichlet algebras were introduced by J. Wermer [2] and logmodular algebras by K. Hoffman [1].

Theorem 6.1 was proved by K. Hoffman [1] while proof of Theorem 6.4 is due to J. Wermer [2] for Dirichlet algebras and to K. Hoffman [1] for log-modular algebras. Theorem 6.6 was given by I. Glicksberg, J. Wermer [1] but its most general form belongs to I. Glicksberg [3].

The results of paragraph 6.2 are mainly due to I. Suciu [1], [2], [3], [5]. Proposition 6.13 belongs to I. Glicksberg [1] and Theorem 6.17 to N. Mochizuki [1]. In connection with these results we also mention E. A. Gorin [2].

Theorem 6.20 appears in a first form in H. Helson, F. Quigley [1]. The form under which it is presented here can be found in K. Hoffman, I. M. Singer [1]. Theorems 6.20 and 6.21 are proved by H. S. Bear [1]. Concerning the maximality property see E. A. Gorin [1].

The use of potential theory in approximation problems on compact sets of the complex plane is quite old (cf. M. Brelot [1]). The treatment of paragraph 6.23 follows E. Bishop [3], L. Carleson [2] and I. Glicksberg, J. Wermer [1].

Proposition 6.29 and Theorem 6.39 belong to D. R. Wilken [1].

The proof of Proposition 6.34 belongs to I. Glicksberg, J. Wermer [1] as well as the proof of Mergelyan theorem [1] (Theorem 6.35).

Proposition 6.37 was given by J. Wermer [1], Proposition 6.39 by H. Helson [1] and Proposition 6.40 by K. Hoffman [3].

In connection with these problems we also mention L. Carleson [1], D. Gaşpar [1], S. Merrill [1], N. Mochizuki [2], [3].

CHAPTER 7

Operator representations of function algebras

7.1. Positive definite maps on $C(X)$. Spectral and semispectral measures

Let X be a compact Hausdorff space and H a Hilbert space. We denote by $L(H)$ the algebra of all bounded linear operators on H.

The linear map $\varphi \to T_\varphi$ of $C(X)$ in $L(H)$ is called *positive definite* if for any $\varphi \in C(X)$, $\varphi \geqslant 0$, we have $T_\varphi \geqslant 0$.

One easily verifies that if $\varphi \to T_\varphi$ is positive definite then it is symmetric ($T_{\overline{\varphi}} = T_\varphi^*$, where $\overline{\varphi}$ is the complex conjugate of φ) and continuous ($\|T_\varphi\| \leqslant K \|\varphi\|$).

For $h, k \in H$ we put

$$\mu_{h,k}(\varphi) = (T_\varphi h, k) \qquad (\varphi \in C(X)). \tag{7.1.1}$$

Thus, we obtain a family $(\mu_{h,k})_{h,k \in H}$ of measures on X, with the following properties:

$$\begin{aligned}
&\mu_{\lambda_1 h_1 + \lambda_2 h_2, k} = \lambda_1 \mu_{h_1,k} + \lambda_2 \mu_{h_2,k}\\
&\mu_{h,k}(\varphi) = \overline{\mu_{k,h}(\overline{\varphi})}\\
&\mu_{h,h} \geqslant 0\\
&\mu_{h,k} \leqslant K \|h\| \, \|k\|.
\end{aligned} \tag{7.1.2}$$

A family $(\mu_{h,k})_{h,k\in H}$ of measures on X is called *semispectral* if it satisfies relations (7.1.2).

We shall write $\mu_h = \mu_{h,h}$. It is easy to see that

$$4\mu_{h,k} = \mu_{h+k} - \mu_{h-k} + i\mu_{h+ik} - i\mu_{h-ik} \qquad (h, k \in H) \tag{7.1.3}$$

and

$$\mu_{h+k} + \mu_{h-k} = 2\mu_h + 2\mu_k \qquad (h, k \in H). \tag{7.1.4}$$

Theorem 7.1. *Relation (7.1.1) gives a one-to-one correspondence between the set of all positive definite maps $\varphi \to T_\varphi$ of $C(X)$ in $L(H)$ and the set of all semispectral families $(\mu_{h,k})_{h,k\in H}$ on X.*

Proof. If $\varphi \to T_\varphi$ is a positive definite map then, as we have already seen, (7.1.1) defines a semispectral family on X with values in H. If $(\mu_{h,k})_{h,k\in H}$ is a semispectral family on X, then using (7.1.1) we may construct the operator T_φ for any $\varphi \in C(X)$; by a simple calculation we obtain that $\varphi \to T_\varphi$ is a positive definite map of $C(X)$ in $L(H)$. ◇

Let us denote by $B(X)$ the set of Borel parts of X. A map $\sigma \to F(\sigma)$ of $B(X)$ in $L(H)$ is called a *semispectral measure* (on X with values in $L(H)$) if for any $h \in H$ the map $\sigma \to (F(\sigma)h, h)$ is a positive Borel measure on X. A semispectral measure on X with values in $L(H)$ will be denoted by $(F(\sigma))_{\sigma\in B(X)}$.

Theorem 7.2. *There exists a one-to-one correspondence between the set of all semispectral families $(\mu_{h,k})_{h,k\in H}$ and the set of all semispectral measures $(F(\sigma))_{\sigma\in B(X)}$, given by*

$$\mu_h(\sigma) = (F(\sigma)h, h). \tag{7.1.5}$$

Proof. If $(\mu_{h,k})_{h,k\in H}$ is a semispectral family on X then, it is clear that, if we put

$$(F(\sigma)h, k) = \mu_{h,k}(\sigma) \qquad (\sigma \in B(X);\ h, k \in H)$$

we can construct the semispectral measure $(F(\sigma))_{\sigma\in B(X)}$ which satisfies (7.1.5).

If $(F(\sigma))_{\sigma\in B(X)}$ is a semispectral measure on X, then for any $h \in H$, μ_h defined by (7.1.5) is a Borel measure on X. (7.1.4) is verified by a simple calculus. Now, defining $\mu_{h,k}$ for any $h, k \in H$ by (7.1.3), one easily obtains that $(\mu_{h,k})_{h,k\in H}$ is a semispectral family on X.

The 3-tuple $\varphi \to T_\varphi$ — a positive definite map of $C(X)$ in $L(H)$, $(\mu_{h,k})_{h,k\in H}$ —a semispectral family, $(F(\sigma))_{\sigma\in B(X)}$ —the semispectral measure, is therefore well defined in the sense of Theorems 7.1 and 7.2, by one of its terms.

The semispectral measure $(E(\sigma))_{\sigma\in B(X)}$ is called *spectral* if for any $\sigma_1, \sigma_2 \in B(X)$ we have $E(\sigma_1 \cap \sigma_2) = E(\sigma_1)\, E(\sigma_2)$.

Theorem 7.3. *Let $\varphi \to T_\varphi$ be a positive definite map and $(F(\sigma))_{\sigma\in B(X)}$ its semispectral measure. The measure $(F(\sigma))_{\sigma\in B(X)}$ is spectral if and only if $\varphi \to T_\varphi$ is a multiplicative map.*

Proof. Let $\varphi \to T_\varphi$ be a positive definite map of $C(X)$ in $L(H)$, $(\mu_{h,k})_{h,k\in H}$ be the semispectral family attached to $\varphi\to T_\varphi$, and $(F(\sigma))_{\sigma\in B(X)}$ its semispectral measure.

Suppose that $\varphi \to T_\varphi$ is multiplicative. For any $f, g \in C(X)$ and $h, k \in H$ we have

$$\int fg\mathrm{d}\mu_{h,k} = (T_{fg}h, k) = (T_fT_g\, h, k) = \int f\mathrm{d}\mu_{T_gh,\,k}$$

$$\int f\mathrm{d}\mu_{T_gh,k} = (T_fT_g\, h, k) = (T_gT_f\, h, k) = (T_fh,\, T_{\bar g}k) = \int f\mathrm{d}\mu_{h,\,T_{\bar g}k}.$$

We therefore have the following measure equalities

$$\text{(7.1.6)}\qquad \begin{aligned} g\mathrm{d}\mu_{h,k} &= \mathrm{d}\mu_{T_gh,k} \\ \mathrm{d}\mu_{T_gh,k} &= \mathrm{d}\mu_{h,\,T_{\bar g}k} \end{aligned} \qquad (g\in C(X);\, h, k\in H).$$

Using (7.1.6). for any $\sigma\in B(X)$, $g \in C(X)$ and $h, k \in H$, we have

$$\int g\chi_\sigma\mathrm{d}\mu_{h,k} = \int \chi_\sigma\mathrm{d}\mu_{T_gh,k}$$

$$\text{(7.1.7)}\qquad \int g\mathrm{d}\mu_{F(\sigma)h,k} = (T_gF(\sigma)h, k) = (F(\sigma)h,\, T_{\bar g}k) =$$

$$= \int \chi_\sigma\mathrm{d}\mu_{h,\,T_{\bar g}k} = \int\chi_\sigma\, \mathrm{d}\mu_{T_gh,k}.$$

We have the measure equality

$$\chi_\sigma \mathrm{d}\mu_{h,k} = \mathrm{d}\mu_{F(\sigma)h,k} \qquad (\sigma \in B(X);\, h, k \in H) \tag{7.1.8}$$

If we use (7.1.8) for any $\sigma, \omega \in B(X)$, we get

$$(F(\sigma \cap \omega)h, k) = \mu_{h,k}(\sigma \cap \omega) = \int \chi_\sigma \chi_\omega \, \mathrm{d}\mu_{h,k} =$$

$$= \int \chi_\sigma \, \mathrm{d}\mu_{F(\omega)h,k} = \int \mathrm{d}\mu_{F(\sigma)F(\omega)h,k} =$$

$$= (F(\sigma)F(\omega)h, k)$$

that is

$$F(\sigma \cap \omega) = F(\sigma)F(\omega) \qquad (\sigma, \omega \in B(X))$$

and therefore $(F(\sigma))_{\sigma \in B(X)}$ is a spectral measure.

Conversely, assume $(F(\sigma))_{\sigma \in B(X)}$ a spectral measure on X. As for any $\sigma \in B(X)$, $F(\sigma)$ is a projection (selfadjoint) of $L(H)$, then for any $\sigma, \omega \in B(X)$ we have

$$\int \chi_\omega \mathrm{d}\mu_{F(\sigma)h,k} = (F(\omega)F(\sigma)h, k) = (F(\omega \cap \sigma)h, k) =$$

$$= \int \chi_\omega \chi_\sigma \, \mathrm{d}\mu_{h,k}$$

$$\int \chi_\omega \mathrm{d}\mu_{F(\sigma)h,k} = (F(\omega)\, F(\sigma)h, k) = (F(\omega)h, F(\sigma)k) =$$

$$= \int \chi_\omega \mathrm{d}\mu_{h,F(\sigma)k}.$$

We therefore obtain the following measure equalities

$$\begin{aligned} &\chi_\sigma \mathrm{d}\mu_{h,k} = \mathrm{d}\mu_{F(\sigma)h,k} \\ &\mathrm{d}\mu_{F(\sigma)h,k} = \mathrm{d}\mu_{h,F(\sigma)k} \end{aligned} \qquad (\sigma \in B(X);\, h, k \in H). \tag{7.1.9}$$

According to (7.1.9), for any $\sigma \in B(X)$ and $g \in C(X)$ we have

$$\int g\chi_\sigma \, d\mu_{h,k} = \int g d\mu_{F(\sigma)h,k}$$

$$\int \chi_\sigma d\mu_{T_g h,k} = \overline{\int \chi_\sigma d\mu_{k,T_g h}} = \overline{\int d\mu_{F(\sigma)k,T_g h}} =$$

$$= \overline{(F(\sigma)k, T_g h)} = (T_g h, F(\sigma)k) =$$

$$= \int g d\mu_{h,F(\sigma)k} = \int g d\mu_{F(\sigma)h,k}.$$

Hence there results

$$g d\mu_{h,k} = d\mu_{T_g h,k}. \tag{7.1.10}$$

From (7.1.10) we obtain

$$(T_f T_g h, k) = \int f d\mu_{T_g h,k} = \int f g d\mu_{h,k} = (T_{fg} h, k)$$

which proves that $\varphi \to T_\varphi$ is multiplicative. ◇

In general, we denote by $\varphi \to U_\varphi$ the multiplicative maps and by $(E(\sigma))_{\sigma \in B(X)}$ the spectral measures. A semispectral family attached to a multiplicative map (or to a spectral measure) will be called a *spectral family*.

The notion of positive definite map introduced above is justified by the following

Proposition 7.4. *Let $\varphi \to T_\varphi$ be a positive definite map of $C(X)$ in $L(H)$. For any finite system $g_1, \ldots, g_n$ of functions in $C(X)$ and any finite system $h_1, \ldots, h_n$ of elements in H, we have*

$$\sum_{i,j} (T_{\bar{g}_i g_j} h_j, h_i) \geqslant 0. \tag{7.1.11}$$

Proof. Let $(\mu_{h,k})_{h,k \in H}$ be the semispectral family attached to the map $\varphi \to T_\varphi$. For the finite system $h_1, \ldots, h_n$ of elements in H we define the positive measure μ on X by

$$\mu = \sum_{i=1}^{n} \mu_{h_i}.$$

Following (7.1.3) and (7.1.4), for any i, j, the measure μ_{h_i, h_j} is absolutely continuous with respect to μ. Hence

$$d\mu_{h_i, h_j} = f_{ij} d\mu$$

with $f_{ij} \in L^1(d\mu)$. For any $\varphi \in C(X)$ and any finite system $\lambda_1, \dots \lambda_n$ of complex numbers, we have

$$\int \varphi \sum_{i,j} \lambda_i \bar{\lambda}_j f_{ij} d\mu = \sum_{i,j} \lambda_i \bar{\lambda}_j \int \varphi d\mu_{h_j, h_i} =$$

$$= \sum_{i,j} \bar{\lambda}_j \lambda_i (T_\varphi h_j, h_i) = (T_\varphi \sum_j \bar{\lambda}_j h_j, \sum_i \bar{\lambda}_i h_i).$$

Therefore, for any $\varphi \geqslant 0$

$$\int \varphi \sum_{i,j} \lambda_i \bar{\lambda}_j f_{ij} d\mu \geqslant 0.$$

Thus, for any countable family $\mathscr{F}$ of finite systems $\lambda_1, \dots, \lambda_n$ of complex numbers, we find a Borel set ω such that $\mu(X - \omega) = 0$ and

$$\sum_{i,j} \lambda_i \bar{\lambda}_j f_{ij}(x) \geqslant 0$$

for $x \in \omega$ and $\lambda_1, \dots, \lambda_j$ in $\mathscr{F}$. If, as $\mathscr{F}$, we take the set of all systems $\lambda_1, \dots, \lambda_n$ for which λ_i, $1 \leqslant i \leqslant n$, has rational real and imaginary parts and we consider the corresponding ω then, for any $x \in \omega$ the numerical matrix $(f_{ij}(x))$ is positive definite.

Let now $g_1, \dots, g_n$ be a system of functions in $C(X)$. Since for any system $\lambda_1, \dots, \lambda_n$ of complex numbers and any $x \in X$ we have

$$\sum_{i,j} \lambda_i \bar{\lambda}_j \bar{g}_i g_j = | \sum_j \overline{\lambda_j g_j} |^2 \geqslant 0$$

then the matrix $(\bar{g}_i g_j(x))$ is positive definite for any $x \in X$. Therefore we clearly have

(7.1.12) $$\sum_{i,j} \bar{g}_i(x) g_j(x) f_{ij}(x) \geqslant 0 \qquad (x \in \omega).$$

Using (7.1.12) we get

$$\sum_{i,j}(T_{\bar{g}_i g_j}h_j, h_i) = \sum_{i,j}\int \bar{g}_i g_j \mathrm{d}\mu_{h_j,h_i} = \sum_{i,j}\int \bar{g}_i g_j f_{ij}\mathrm{d}\mu =$$

$$= \int \sum_{i,j}\bar{g}_i g_j f_{ij}\mathrm{d}\mu = \int_{\omega} \sum_{i,j}\bar{g}_i g_j f_{ij}\mathrm{d}\mu \geqslant 0.$$

The proposition is proved. ◇

Theorem 7.5. (The dilatation theorem of M. A. Naimark) *Let $\varphi \to T_\varphi$ be a positive definite map of $C(X)$ in $L(H)$. There exists a Hilbert space K, a bounded linear operator $V\colon H \to K$ and a multiplicative linear map $\varphi \to U_\varphi$ of $C(X)$ in $L(K)$, which satisfy $U_1 = I$ — the identity operator of K, $\|U_\varphi\| \leqslant \|\varphi\|$, $\varphi \in C(X)$ and*

$$T_\varphi = V^* U_\varphi V \qquad (\varphi \in C(X)). \tag{7.1.13}$$

Proof. Consider the tensor product $C(X)\otimes H$, with elements of the form $u = \sum_j f_i \otimes h_j$, where $f_i \in C(X)$, $h_j \in H$, and the sum having a finite number of non-zero terms. We define the sesquilinear form

$$\langle u, v\rangle = \sum_{i,j}(T_{\bar{g}_i f_j}h_j, k_i) \tag{7.1.14}$$

if $u = \sum_j f_j \otimes h_j$, $v = \sum_i g_i \otimes k_i$.

Since $\varphi \to T_\varphi$ is positive definite, from Proposition 7.4 there results

$$\langle u, u\rangle = \sum_{i,j}(T_{\bar{f}_i f_j}h_j, h_i) \geqslant 0$$

hence $\langle u, v\rangle$ is a positive definite form.

Let $N = \{u \in C(X)\otimes H\colon \langle u, u\rangle = 0\}$. The form $\langle u, v\rangle$ induces an inner product on the quotient space $C(X)\otimes H/N$. K will be the Hilbert space obtained by completing $C(X)\otimes H/N$ with respect to this inner product.

We now define the operator V on H, with values in K, by

$$Vh = 1 \otimes h + N.$$

It is clear that V is linear and

$$\|Vh\|^2 = (Vh, Vh) = \langle 1 \otimes h, 1 \otimes h\rangle = (T_1 h, h) \leqslant \|T_1\| \, \|h\|^2.$$

Hence V is bounded.

Let us define the map $\varphi \to U_\varphi$. We first define it for $u = \sum_j g_j \otimes h_j$ and $\varphi \in C(X)$.

$$U_\varphi^0\left(\sum_j g_j \otimes h_j\right) = \sum_j \varphi g_j \otimes h_j .$$

It is immediate that $\varphi \to U_\varphi^0$ is linear and multiplicative. Moreover

$$(U_\psi^0 u, u) = \left(\sum_j \varphi g_j \otimes h_j, \sum_i g_i \otimes h_i\right) = \sum_{i,\,l} (T_{\bar{g}_i \varphi g_j} h_j, h_i) =$$

$$= \sum_{l,\,j} (T_{\overline{\bar{\varphi} g_i} g_j} h_j, h_i) = \left(\sum_j g_j \otimes h_j, \sum_l \bar{\varphi}_{g_l} \otimes h_i\right) =$$

$$= (u, U_\varphi^0 u).$$

Hence

$$(U_\varphi^0 u, u) = (u, U_{\bar{\varphi}}^0 u) \qquad (\varphi \in C(X)). \tag{7.1.15}$$

For a fixed u we define, on $C(X)$, the functional

$$\rho(\varphi) = (U_\varphi^0 u, u) \qquad (\varphi \in C(X)).$$

ρ is clearly a linear functional on $C(X)$. Since for any $\varphi \in C(X)$, using (7.1.5), we have

$$\rho(\varphi\bar{\varphi}) = (U_{\varphi\bar{\varphi}}^0 u, u) = (U_\varphi^0 u, U_\varphi^0 u) \geqslant 0$$

ρ is a positive functional and $\|\rho\| \leqslant \rho(1) = \langle u, u\rangle$. Then

$$\langle U_\varphi^0 u, U_\varphi^0 u\rangle = \langle U_{\varphi\bar{\varphi}}^0 u, u\rangle = \rho(\varphi\bar{\varphi}) \leqslant \|\varphi\|^2 \langle u, u\rangle.$$

Now, if we put

$$U_\varphi(u + N) = U^0_\varphi u + N$$

and extended continuously on all K, then we clearly obtain a multiplicative linear map $\varphi \to U_\varphi$ of $C(X)$ in $L(K)$, which satisfies $U_1 = I$, $\|U_\varphi\| \leqslant \|\varphi\|$, $\varphi \in C(X)$.

Moreover, for any $h, k \in H$, $\varphi \in C(X)$, we have

$$(V^* U_\varphi Vh, k) = (U_\varphi Vh, Vk) = (U_\varphi 1 \otimes h, 1 \otimes k) =$$

$$= (\varphi \otimes h, 1 \otimes k) = (T_\varphi h, k).$$

Hence

$$T_\varphi = V^* U_\varphi V \qquad (\varphi \in C(X)).$$

The theorem is proved. ◇

Let us note that if $T_1 = I_H$ — the identity operator on H, then $V^*V = I_K$ — the identity operator on K, that is V is an isometry. At the same time $P = VV^*$ is the orthogonal projection of K on VH. Then, identifying H with VH, we may consider H as a subspace of K, hence

$$T_\varphi h = PU_\varphi h \qquad (\varphi \in C(X);\, h \in H).$$

To complete the dilation theorem of M. A. Naimark we now give the following uniqueness theorem.

Theorem 7.6. *Let $\varphi \to T_\varphi$ be a positive definite map of $C(X)$ in $L(H)$ and K_i, V_i, $\varphi \to U^i_\varphi$, $i = 1, 2$, be as in Theorem 7.5. Moreover, we assume that the elements $U^i_\varphi V_i\, h$, $\varphi \in C(X)$, $h \in H$, span the space K_i. Then there exists a unitary operator $U\colon K_2 \to K_1$ such that*

$$UV_2 = V_1$$

$$U^2_\varphi = U^* U^1_\varphi U \qquad (\varphi \in C(X)).$$

Proof. We define, on K_2, the operator U by

$$U(\sum_i U^2_{g_i} V^2 h_i) = \sum_i U^1_{g_i} V^1 h_i.$$

Then

$$\|U(\sum_j U^2_{g_j}V_2h_j\|^2 = (U(\sum_j U^2_{g_j}V_2h_j), U(\sum_i U^2_{g_i}V_2h_i)) =$$

$$= (\sum_j U^1_{g_j}V_1h_j, \sum_i U^1_{g_i}V_1h_i) = \sum_{i,j} (V_1^* U_{\bar{g}_i g_j}V_1h_j, h_j) =$$

$$= \sum_{i,j} (T_{\bar{g}_i g_j}h_i, h_j) = \sum_{i,j} (V_2^* U^2_{\bar{g}_i g_j}V_2h_i, h_j) = \ldots =$$

$$= \|\sum_j U^2_{g_j}V_2\, h_j\|^2.$$

Now it is clear that U may be defined on all K_2 and that it is a unitary operator. It follows immediately from the definition of U that it satisfies the conditions required by the theorem.

7.2. Representations of function algebras

Let X be a compact Hausdorff space, A a function algebra on X, H a complex Hilbert space and $L(H)$ the algebra of all bounded linear operators on H.

An algebra homomorphism $f \to T_f$ of A in $L(H)$, which satisfies

$$T_1 = I \tag{7.2.1}$$

and

$$\|T_f\| \leqslant \|f\| \tag{7.2.2}$$

is called a *representation* of A on H.

Proposition 7.7. *Let $f \to T_f$ be a representation of A on H. If f and $\bar{f} \in A$ then $T_{\bar{f}} = T_f^*$.*

Proof. It is sufficient to prove that if f is a real function in A then $T_f = T_f^*$. Indeed, if $f, \bar{f} \in A$, then the real functions $u = \frac{1}{2}(f+\bar{f})$, $v = \frac{i}{2}(f-\bar{f})$ belong to A and, since $f = u + iv$, we have

$$T_f = T_u + iT_v, \quad T_{\bar{f}} = T_u - iT_v.$$

But, as $T_u = T_u^*$, $T = T_v^*$, there follows

$$T_f^* = T_u^* - iT_v^* = T_u - iT_v = T_{\bar{f}}.$$

Thus, let f be a real function in A. Let t be a real number, $t > \|f\|$. Then the functions $(f \pm it)^{-1}$ belong to A. Write

$$g_1 = (f + it)(f - it)^{-1};\ g_2 = (f - it)(f + it)^{-1}.$$

We have $g_1, g_2 \in A$, $g_1 g_2 = 1$, $|g_1| = |g_2| = 1$. Hence $T_{g_1}^{-1}$ exists and $T_{g_1}^{-1} = T_{g_2}$. As $\|T_{g_1}\| \leqslant \|g_1\| = 1$, $\|T_{g_1}^{-1}\| = \|T_{g_2}\| \leqslant \|g_2\| = 1$, T_{g_1} results a unitary operator. But

$$T_{g_1}(T_f - itI) = T_f + itI \tag{7.2.3}$$

and therefore we have

$$4\mathrm{Re}\,it(T_f h, h) = \|(T_f - itI)h\|^2 - \|(T_f + itI)h\|^2 =$$

$$= \|(T_f - itI)h\|^2 - \|T_{g_1}(T_f - itI)h\|^2 = 0$$

for any $h \in H$. Here we have a well-known relation in the geometry of Hilbert spaces, relation (7.2.3) and the fact that T_{g_1} is unitary. Then $(T_f h, h)$ is real for any $h \in H$, hence $T_f^* = T_f$. ◇

Proposition 7.8. *Let* $f \to T_f$ *be a representation of* A *on* H. *If* $\mathrm{Re} f \geqslant 0$ *then* $\mathrm{Re}\,T_f \geqslant 0$.

Proof. If $\mathrm{Re} f \geqslant 0$, then $(f + 1)^{-1} \in A$. Let $G = (f - 1)(f + 1)^{-1}$; we have $g \in A$, $\|g\| \leqslant 1$, hence $\|T_g\| \leqslant 1$. Since

$$T_g(T_f + I) = T_f - I$$

for any $h \in H$ we get

$$4\mathrm{Re}(T_f h, h) = \|(T_f + I)h\|^2 - \|T_f - I)h\|^2 =$$

$$= \|(T_f + I)h\|^2 - \|T_g(T_f + I)h\|^2 \geqslant 0.$$

Therefore

$$\mathrm{Re}T_f = \frac{1}{2}(T_f + T_f^*) \geqslant 0. \ \diamond$$

Corollary 7.9. *For any $f \in A$ we have*

$$\|\mathrm{Re}T_f\| \leqslant \|\mathrm{Re}f\|.$$

Corollary 7.10. *A representation $\varphi \to U_\varphi$ of $C(X)$ on H is a positive definite multiplicative map of $C(X)$ in $L(H)$.*

7.3. Representations of the algebra $C(X)$

We have seen that any representation of the algebra $C(X)$ on a Hilbert space H is a positive definite multiplicative map $\varphi \to U_\varphi$ of $C(X)$ in $L(H)$. Hence there exists a spectral measure $(E(\sigma))_{\sigma \in B(X)}$ on X with values in $L(H)$, such that

$$(U_\varphi h, k) = \int_X \varphi(x)\mathrm{d}(E(x)h, k) \qquad (h, k \in H;\ \varphi \in C(X)).$$

In this paragraph we shall describe completely the form of the representations of the algebra $C(X)$ on a separable Hilbert space H.

Let $\varphi \to U_\varphi$ be a representation of $C(X)$ on a Hilbert space H. A closed subspace M of H will be called *cyclic* (for the representation $\varphi \to U_\varphi$) if there exists $h \in M$ such that the closure M_h of the set $\{U_\varphi h: \varphi \in C(X)\}$ be equal to M. For any $h \in H$, M_h is obviously a cyclic subspace, called a *cyclic subspace generated by h.*

Two non-zero elements $h_1, h_2 \in H$ are said to be *cyclic free* if $M_{h_1} \perp M_{h_2}$. A non-void system S of non-zero elements of H is called *cyclic free* if it reduces to one element or, if any two elements of S are cyclic free.

One easily verifies that the set of cyclic free systems of H is inductively ordered, hence any element of H belongs to a maximal cyclic free system.

Proposition 7.11. *Let S be a maximal cyclic free system of H. Then*

$$H = \bigoplus_{h \in S} M_h.$$

Proof. According to the definition of a cyclic free system, $M_{h_1} \perp M_{h_2}$ for any $h_1, h_2 \in S$, $h_1 \neq h_2$. Let now $h_1 \in H$ be orthogonal on any M_h, $h \in S$. We then have

$$(U_{\varphi_1}h, U_{\varphi_2}h_1) = (U_{\varphi_1\bar{\varphi}_2}h, h_1) = 0 \qquad (\varphi_1, \varphi_2 \in C(X)).$$

There results $M_{h_1} \perp M_h$ for any $h \in S$. If $h_1 \neq 0$ then $S \cup \{h_1\}$ is a cyclic free system which is in contradiction with the maximality of S. Therefore $h_1 = 0$, hence

$$H = \bigoplus_{h \in S} M_h. \quad \diamond$$

Let μ be a positive measure on X and N be a Hilbert space. It is already known how to construct the Hilbert space $L^2(N; \mathrm{d}\mu)$ of all functions h defined on X, with values in N, for which

$$\int \|h(x)\|^2 \mathrm{d}\mu(x) < \infty.$$

The representation $\varphi \to U_\varphi$ of $C(X)$ on $L^2(N; \mathrm{d}\mu)$, where

$$U_\varphi h = \varphi h \qquad (h \in L^2(N; \mathrm{d}\mu))$$

will be called the *natural representation of $C(X)$ on $L^2(N; \mathrm{d}\mu)$.*

Theorem 7.12. *Let $\varphi \to U_\varphi$ be a representation of $C(X)$ on a separable Hilbert space H. There exists a positive measure μ on X, a partition $X = \bigcup_{k=1,2,\ldots,\infty} \omega_k$ of X, of disjoint Borel sets and, for any $k = 1,\ldots, \infty$ a k-dimensional Hilbert space N_k such that H is isometrically isomorphic to $\bigoplus_{k=1,2,\ldots,\infty} L^2(N_k; \chi_{\omega_k} \mathrm{d}\mu)$ and the representation $\varphi \to U_\varphi$ is unitary equivalent to the direct sum of the natural representation of $C(X)$ on $L^2(N_k; \chi_{\omega_k}\mathrm{d}\mu)$.*

Proof. Let S be a maximal cyclic free system of elements in H. Since

$$H = \bigoplus_{h \in S} M_h$$

and H is separable, the set S is at most countable. Let $S = \{h_1, h_2, \ldots\}$. We have

$$H = \bigoplus_{i=1}^{\infty} M_{h_i}.$$

We can assume $\|h_i\| = 1$.

For $i = 1, 2, \ldots$ let

$$\mu_i(\varphi) = (U_\varphi h_i, h_i) \qquad (\varphi \in C(X)).$$

μ_i is known to be a positive measure on X and, since $\mu_i(1) = \|h_i\|^2 = 1$, there results $\|\mu_i\| = 1$.

Consider

$$\mu = \sum_{i=1}^{\infty} \frac{1}{2^i} \mu_i.$$

Then μ is a positive measure on X and $\mu(1) = 1$. The measures μ_i are absolutely continuous with respect to μ.

Let $d\mu_i = f_i d\mu$ with $f_i \in L^1(d\mu)$, $f_i \geqslant 0$. Let us denote by $S_i = \{x \in X : f_i(x) \neq 0\}$. For any set of natural numbers $\mathcal{J}$ we put $E_{\mathcal{J}} = (\bigcap_{i \in \mathcal{J}} S_i) - (\bigcup_{i \notin \mathcal{J}} S_i)$. We have $E_{\mathcal{J}_1} \cap E_{\mathcal{J}_2} = \varnothing$ if $\mathcal{J}_1 \neq \mathcal{J}_2$. For any $k = 1, 2, \ldots, \infty$ we write

$$\omega_k = \bigcup_{\text{card } \mathcal{J} = k} E_{\mathcal{J}}.$$

It is clear that a point $x \in X$ belongs to ω_k, $k < \infty$, if there are exactly k sets $S_{i_1}, \ldots, S_{i_k}$ defined as above, which contain x.

We have $\omega_i \cap \omega_j = \varnothing$, $i, j = 1, 2, \ldots, \infty$; $i \neq j$ and

$$X = \bigcup_{k=1,2,\ldots,\infty} \omega_k.$$

Since S_i are countable sets and, since by all operations from (S_i) to (ω_k) for any finite k, we still remain in the Borel clan of measurable sets, ω_k are measurable.

Since

$$\omega_\infty = X - \left(\bigcup_{k=1}^{\infty} \omega_k\right),$$

ω_∞ is measurable too.

As we shall see, the sets ω_k may be chosen to be Borel sets.

For any $k = 1, 2, \ldots, \infty$, let N_k be a Hilbert space of dimension k $(N_\infty = l^2)$.

Now let $h \in H$ have the form $h = \sum_{i=1}^{\infty} U_{\varphi_i} h_i$.

For any $k = 1, 2, \ldots$ we define the function g_h^k on X, with values in N_k, by

$$g_h^k(x) = \begin{cases} (\sqrt{f_{j_1}(x)}\varphi_{j_1}(x), \sqrt{f_{j_2}(x)}\varphi_{j_2}(x), \ldots, \sqrt{f_{j_k}(x)}\varphi_{j_k}(x)) \\ \text{if there exists } \mathscr{J} \text{ such that } x \in E_{\mathscr{J}}, \text{ where} \\ \mathscr{J} = (j_1, j_2, \ldots, j_k), j_1 < j_2 < \ldots < j_k. \\ 0 \text{ for the other } x \end{cases}$$

For $k = \infty$, $\mathscr{J} = (j_1, j_2, \ldots), j_1 < j_2 < \ldots$.

We now show that $g_h^k \in L^2(N_k; \chi_{\omega_k} \mathrm{d}\mu)$. We have

$$\int_X \|g_h^k(x)\|^2 \chi_{\omega_k} \mathrm{d}\mu = \int_{\omega_k} \|g_h^k(x)\|^2 \mathrm{d}\mu = \sum_{\mathrm{card}\, \mathscr{J} = k} \int_{E_{\mathscr{J}}} \|g_h^k(x)\|^2 \mathrm{d}\mu =$$

$$= \sum_{\mathrm{card}\, \mathscr{J} = k} \int_{E_{\mathscr{J}}} (f_{j_1}(x)|\varphi_{j_1}(x)|^2 + f_{j_2}(x)|\varphi_{j_2}(x)|^2 + \ldots + f_{j_k}(x)|\varphi_{j_k}(x)|^2 \mathrm{d}\mu(x) \leqslant$$

$$\leqslant \sum_{i=1}^{\infty} \int |\varphi_i(x)|^2 f_i(x) \mathrm{d}\mu(x) = \sum_{i=1}^{\infty} \int_X |\varphi_i(x)|^2 \mathrm{d}\mu_i =$$

$$= \sum_{i=1}^{\infty} (U_{|\varphi_i|^2} h_i, h_i) = \sum_{i=1}^{\infty} (U_{\varphi_i} h_i, U_{\varphi_i} h_i) = \sum_{i=1}^{\infty} \|U_{\varphi_i} h_i\|^2 = \|h\|^2.$$

Let g_h be the element of $\bigoplus_{k=1,2,\ldots,\infty} L^2(N_k, \chi_{\omega_k} d\mu)$ defined by

$$g_h = (g_h^1, g_h^2, \ldots, g_h^\infty).$$

For $h \in H$ of form $h = \sum_{i=1}^{\infty} U_{\varphi_i} h_i$, we put

$$Uh = g_h.$$

One easily verifies that U is a linear map from the subspace of elements $\sum_{i=1}^{\infty} U_{\varphi_i} h_i$ of H, which is dense in H, into $\bigoplus_{k=1,2,\ldots,\infty} L^2(N_k, \chi_{\omega_k} d\mu)$. Moreover

$$\|Uh\|^2 = \|g_h\|^2 = \sum_{k=1,2,\ldots,\infty} \|g_h^k\|^2 = \sum_{k=1,2,\ldots,\infty} \int_{\omega_k} \|g_h^k(x)\|^2 d\mu =$$

$$= \sum_{k=1,2,\ldots,\infty} \sum_{\operatorname{card} \mathscr{J}=k} \sum_{i=1}^{k} \int_{E_{\mathscr{J}}} |\varphi_{j_i}(x)|^2 f_{j_i} d\mu =$$

$$= \sum_{k=1,2,\ldots,\infty} \sum_{\operatorname{card} \mathscr{J}=k} \sum_{i=1}^{\infty} \int_{E_{\mathscr{J}}} |\varphi_i(x)|^2 f_i(x) d\mu =$$

$$= \sum_{i=1}^{\infty} \sum_{k=1,2,\ldots,\infty} \sum_{\operatorname{card} \mathscr{J}=k} \int_{E_{\mathscr{J}}} |\varphi_i(x)|^2 f_i(x)\, d\mu =$$

$$= \sum_{i=1}^{\infty} \sum_{k=1,2,\ldots,\infty} \int |\varphi_i(x)|^2 f_i(x)\, d\mu = \sum_{i=1}^{\infty} \int |\varphi_i(x)|^2 f_i(x) d\mu =$$

$$= \sum_{i=1}^{\infty} \int |\varphi_i(x)|^2\, d\mu_i = \sum_{i=1}^{\infty} (U_{\varphi_i \bar{\varphi}_i} h_i, h_i) = \sum_{i=1}^{\infty} (U_{\varphi_i} h_i, U_{\varphi_i} h_i) =$$

$$= \sum_{i=1}^{\infty} \|h_i\|^2 = \|h\|^2.$$

At a certain step we used the fact that for $x \in E_{\mathscr{J}}$ we have $f_i(x) = 0$ for any $i \notin \mathscr{J}$.

The map U can therefore be extended to an isometric operator from H into $\bigoplus_{k=1,2,\ldots,\infty} L^2(N_k;\ \chi_{\omega k} d\mu)$.

We now prove that the image of H by U is dense in $\bigoplus_{k=1,2,\ldots,\infty} L^2(N_k; \chi_{\omega k}\, d\mu)$. Let $g \in \bigoplus_{k=1,2,\ldots,\infty} L^2(N_k; \chi_{\omega k}\, d\mu)$ of the form $g = (g^1, g^2, \ldots, g^\infty)$ with $g^k = (g_1^k, \ldots, g_k^k)$, $g^k \in L^2(N_k;\ \chi_{\omega k}\, d\mu)$, be orthogonal to any Uh with $h \in H$. For $h \in H$ of form $h = \sum_{i=1}^{\infty} U_{\varphi_i} h_i$, we have

$$0 = (Uh, g) = (g_h, g) =$$

$$= \sum_{k=1,2,\ldots,\infty} (g_h^k, g^k) = \sum_{k=1,2,\ldots,\infty} \int (g^k(x), g_h^k(x))\, d\mu =$$

$$= \sum_{k=1,2,\ldots,\infty} \ \sum_{\operatorname{card} \mathscr{J} = k} \ \sum_{i=1}^{k} \int \sqrt{f_{j_i}(x)}\ \varphi_{j_i}(x)\ \overline{g_i^k(x)}\, d\mu.$$

Let now $\varphi \in C(X)$ and n a fixed natural number. If in the last relation we take $\varphi_i = 0$ for $i \neq n$ and $\varphi_n = \varphi$, then we obtain

$$\sum_{k=1,2,\ldots,\infty} \ \sum_{\operatorname{card} \mathscr{J} = k} \ \int_{E_{\mathscr{J}}} \sqrt{f_n(x)}\ \varphi(x)\ \overline{g_n^k(x)}\, d\mu = 0,$$

where we used the fact that $f_n = 0$ on $E_{\mathscr{J}}$ if $n \notin E_{\mathscr{J}}$.

Note that, if $n \in \mathscr{J}$ then $E_{\mathscr{J}} = E_{\mathscr{J}} \cap S_n$ and if $n \notin \mathscr{J}$ then $E_{\mathscr{J}} \cap S_n = \varnothing$. Hence

$$\bigcup_{\operatorname{card} \mathscr{J} = k} E_{\mathscr{J}} = \bigcup_{\operatorname{card} \mathscr{J} = k} (E_{\mathscr{J}} \cap S_n) = \omega_k \cap S_n$$

and therefore

$$\bigcup_{k=1,2,\ldots,\infty} \ \bigcup_{\operatorname{card} \mathscr{J} = k} E_{\mathscr{J}} = \bigcup_{k=1,2,\ldots,\infty} (\omega_k \cap S_n) = S_n.$$

Let F be a function on S_n such that $F = g_n^k$ on $E_{\mathscr{J}}$, if $n \in \mathscr{J}$ and card $\mathscr{J} = k$. From the above considerations there results

$$\int_{S_n} \sqrt{f_n(x)}\ \varphi(x)\ F(x)\, d\mu = 0.$$

Since this equality holds for any $\varphi \in C(X)$, there follows $\sqrt{f_n(x)}\, F(x) = 0$ μ-almost everywhere on S_n and, as $f_n(x) \neq 0$ for $x \in S_n$, we get $F(x) = 0$ for $x \in S_n$, hence $g_h^k(x) = 0$ for $x \in E_{\mathscr{I}}$ with $n \in \mathscr{I}$ and card $\mathscr{I} = k$.

Since n has been arbitrarily taken, there results $g^k = 0$ on any $E_{\mathscr{I}}$ with card $\mathscr{I} = k$, that is $g^k = 0$ on ω_k.

Therefore

$$\|g^k\|^2 = \int_X \|g^k(x)\|^2 \chi_{\omega_k}\, d\mu = 0$$

for any k, i.e. $g^k = 0$, hence $g = 0$.

Then, the operator U is an isometric isomorphism between H and $\bigoplus_{k=1,2,\ldots,\infty} L^2(N_k, \chi_{\omega_k}\, d\mu)$.

Let $\varphi \to U'_\varphi$ be the representation of $C(X)$ on $\bigoplus_{k=1,2,\ldots,\infty} L^2(N_k, \chi_{\omega_k}\, d\mu)$, defined by

$$U'_\varphi = U U_\varphi U^{-1}.$$

If g has the form $g = g_h$, with $h \in H$, $h = \sum_{i=1}^{\infty} U_\varphi h_i$, we have

$$U'_\varphi g = U U_\varphi U^{-1} g_h = U U_\varphi h = U \sum_{i=1}^{\infty} U_{\varphi\varphi_i} h_i = \varphi g_h.$$

As we have seen, the elements of this form are dense in $\bigoplus_{k=1,2,\ldots,\infty} L^2(N_k; \chi_{\omega_k}\, d\mu)$; then

$$U'_\varphi g = \varphi g$$

for any $g \in \bigoplus_{k=1,2,\ldots,\infty} L^2(N_k; \chi_{\omega_k}\, d\mu)$.

The proof is complete. ◇

From the proof of this theorem there also results the following

Proposition 7.13. *Let $\varphi \to U_\varphi$ be a representation of $C(X)$ on H. For any $h \in H$, the representation $\varphi \to U_\varphi | M_h$ of $C(X)$ on M_h is unitarily equivalent to the natural representation of $C(X)$ on $L^2(\mathrm{d}\mu_h)$.*

To conclude this paragraph we shall establish the form of invariant subspaces of natural representations of $C(X)$.

Theorem 7.14. *Let μ be a positive measure on X and $\varphi \to U_\varphi$ the natural representation of $C(X)$ on $L^2(\mathrm{d}\mu)$. The closed subspace M of $L^2(\mathrm{d}\mu)$ is invariant to $\varphi \to U_\varphi$ if and only if M has the form*

$$M = \chi_E L^2(\mathrm{d}\mu)$$

where χ_E is the characteristic function of a Borel set E in $B(X)$. E is μ-essentially uniquely determined by M.

Proof. Let M be a closed subspace of $L^2(\mathrm{d}\mu)$ such that $U_\varphi M \subset M$ for any $\varphi \in C(X)$. Let F be the projection of 1 on M. We have

$$\int \varphi F(1 - \bar{F})\, \mathrm{d}\mu = 0 \qquad (\varphi \in C(X)).$$

Then $F = |F|^2$ μ-almost everywhere, hence $F = \chi_E$ with E a Borel set in X.

It is clear that $\chi_E L^2(\mathrm{d}\mu) \subset M$. Now let h be a function in M, orthogonal to $\chi_E L^2(\mathrm{d}\mu)$. We have

$$\int \varphi h(1 - \chi_E)\, \mathrm{d}\mu = 0$$

and

$$\int \chi_E \varphi h \bar{h}\, \mathrm{d}\mu = 0.$$

Therefore, $h = \chi_E h$ μ-almost everywhere and

$$\int \varphi |h|^2\, \mathrm{d}\mu = \int \varphi h \bar{h}\, \mathrm{d}\mu = \int \chi_E \varphi h \bar{h}\, \mathrm{d}\mu = 0$$

for any $\varphi \in C(X)$. There results $h = 0$ μ-almost everywhere, so $M = \chi_E L^2(\mathrm{d}\mu)$.

If $M = \chi_{E_1} L^2(\mathrm{d}\mu) = \chi_{E_2} L^2(\mathrm{d}\mu)$ with E_1 and E_2 Borel sets in X then, clearly, $\chi_{E_1} = \chi_{E_2}$ μ-almost everywhere.

Notes

A unitary treatment of positive definite maps on $C(X)$, spectral and semi-spectral families and spectral and semispectral measures can be found in C. Foiaş [4] and C. T. Ionescu-Tulcea [1] as well as in M. A. Naimark [3].

The first direct proof to Proposition 7.4 has been given by W. F. Stinespring [1]. Naimark's dilation theorem first appears in M. A. Naimark [2].

A new proof belongs to C. Foiaş [4] and it has also been presented in the Romanian edition of this book. In the present chapter the proof follows W. F. Stinespring [1] where the theorem is proved in the case of C*-algebras. See also W. B. Arveson [2].

A first study of operator representations of arbitrary function algebras has been carried out by C. Foiaş and I. Suciu [1]. The same work contains the results of paragraph 7.2.

The origin of Theorem 7.12 is to be found in the theory of spectral multiplicity (cf. P. R. Halmos [2]); the present proof follows M. Zerner [1].

Another proof of Theorem 7.14 in the case of the unit circle of the complex plane and the Lebesgue measure appears in H. Helson [1].

CHAPTER 8

Elements of spectral theory of representations of function algebras

8.1. The canonical decomposition

Let A be a function algebra on X and H a Hilbert space.

A representation $f \to T_f$ of A on H will be called *X-spectral,* or simply *spectral,* if it is the restriction to A of a representation $\varphi \to U_\varphi$ of $C(X)$ on H. The representation $\varphi \to U_\varphi$ of $C(X)$ on H, which satisfies

$$T_f = U_f$$

is called the *extension* to $C(X)$ of $f \to T_f$.

Proposition 8.1. *A spectral representation $f \to T_f$ of A on H has a unique extension to $C(X)$.*

Proof. Indeed, let $\varphi \to U_\varphi$ and $\varphi \to V_\varphi$ be two extensions to $C(X)$ of the representation $f \to T_f$. Since $U_f = T_f = V_f$ for $f \in A$, there results $U_f^* = T_f^* = V_f^*$, hence

$$U_{\Sigma c_i f_i \bar{g}_i} = \sum c_i U_{f_i} U_{g_i}^* = \sum c_i V_{f_i} V_{g_i}^* = V_{\Sigma c_i f_i \bar{g}_i}$$

for $f_i, g_i \in A$.

Therefore $U_\varphi = V_\varphi$ for any function of the form $\varphi = \Sigma c_i f_i \bar{g}_i$ with $f_i, g_i \in A$. Since the set of all functions with such a form is dense in $C(X)$ and the representations $\varphi \to U_\varphi$, $\varphi \to V_\varphi$ are continuous, there follows $U_\varphi = V_\varphi$ for any $\varphi \in C(X)$.

A closed subspace M of H is said to be *invariant* (to $f \to T_f$) if $T_f M \subset M$ for any $f \in A$. The subspace M is called *doubly invariant* (to $f \to T_f$) if $T_f M \subset M$, $T_f^* M \subset M$ for any $f \in A$.

Proposition 8.2. *Let $f \to T_f$ be a spectral representation of A on H, $\varphi \to U$ its extension to $C(X)$ and M a doubly invariant subspace. Then $U_\varphi M \subset M$ for any $\varphi \in C(X)$.*

Proof. Obviously $U_\varphi M \subset M$ for any φ of the form $\Sigma f_i \bar{g}_i$ with $f_i, g_i \in A$; using the continuity we get then $U_\varphi M \subset M$ for any $\varphi \in (CX)$. ◇

Corollary 8.3. *If $f \to T_f$ is a spectral representation of A on H, $\varphi \to U_\varphi$ its extension to $C(X)$ and M a doubly invariant space, then the representation $f \to T_f | M$ of A on M is spectral and $\varphi \to U_\varphi | M$ is its extension to $C(X)$.*

A closed subspace M of H is called *X-spectral* or, shortly, *spectral* (for $f \to T_f$) if it is doubly invariant and the representation $f \to T_f | M$ of A on M is spectral.

The representation $f \to T_f$ of A on H is called *completely non-spectral* if it has no spectral subspaces different from zero.

Proposition 8.4. *Let M_1 and M_2 be two spectral subspaces for the representation $f \to T_f$ of A on H. Then $M = M_1 + M_2$ is a spectral subspace.*

Proof. Clearly M is doubly invariant and

$$T_f^* T_g h = T_g T_f^* h \qquad (h \in M; f, g \in A). \tag{8.1.1}$$

For finite systems $f_1, \ldots, f_n$, $g_1, \ldots, g_n$ we define the operator

$$T = (\textstyle\sum T_{f_i}^* T_{g_i}) | M. \tag{8.1.2}$$

From (8.1.1) T is a normal operator on M. We shall prove that

$$\|T\| \leqslant \| \textstyle\sum \bar{f}_i g_i \|. \tag{8.1.3}$$

Let $k = \|\Sigma \bar{f}_i g_i\|$ and $E_T(\sigma)$ be the spectral measure of T. To prove (8.1.3) it is sufficient to show that

$$E_T(\{z \in C: |z| \leqslant k\}) M = M,$$

hence it is sufficient to show that

$$M_p \subset E_T(\{z \in C: |z| \leqslant k\}) M \qquad (p = 1, 2).$$

Let $h \in M_p$ and assume $h \notin E_T(\{z \in C: |z| \leqslant k\})\, M$. That means

$$E_T(\{z \in C: |z| > k\})\, h \neq 0$$

hence there exists $\varepsilon > 0$ such that

$$h' = E_T(\{z \in C: |z| \geqslant k + \varepsilon\})\, h \neq 0.$$

We have

$$(k + \varepsilon)^{2^n} \|h'\|^2 = (k + \varepsilon)^{2^n} \int \mathrm{d}(E_T(z)\, h, h) \leqslant$$

$$\leqslant \int\limits_{\{z;\, |z| \geqslant k + \varepsilon\}} |z|^{2^n} \mathrm{d}(E_T(z)\, h, h) \leqslant \int |z|^{2^n} \mathrm{d}(E_T(z)\, h, h) = \|T^{2^n} h\|^2.$$

Then

$$(k + \varepsilon)^{2^n} \|h'\|^2 \leqslant \|T^{2^n} h\|^2. \tag{8.1.4}$$

Let $\varphi \to U^p_\varphi$ be the extension to $C(X)$ of the spectral representation $f \to T_f|M_p$. As $h \in M_h$, we have

$$Th = \sum T^*_{f_i} T_{g_i} h = \sum U^p_{\bar{f}_i} U^p_{g_i} h = U^p_{\Sigma \bar{f}_i g_i} h.$$

Hence

$$T^n h = U_{(\Sigma \bar{f}_i g_i)^n} h$$

and therefore

$$\|T^n h\| = \|U_{(\Sigma \bar{f}_i g_i)^n} h\| \leqslant \|\sum \bar{f}_i g_i\|^n \|h\| = k^n \|h\|. \tag{8.1.5}$$

Using (8.1.5) in (8.1.4) we obtain

$$(k + \varepsilon)^{2^n} \|h'\|^2 \leqslant k^{2^n} \|h\|^2$$

for any natural n, which is impossible. Hence (8.1.3) is true.

Let $\varphi \in C(X)$ be of the form $\varphi = \Sigma \bar{f}_i g_i$ with $f_i, g_i \in A$, and let $U_\varphi = T$, T constructed as above. According to (8.1.3), the map $\varphi \to U_\varphi$ of $C(X)$ in $L(M)$ is well defined and, as it is easily verified, it is a representation of $C(X)$ on M which extends $f \to T_f|M$. Therefore M is spectral, and that is exactly what we had to prove. ◇

Proposition 8.5. *Let $\{M_i\}_{i\in\mathcal{I}}$ be an increasingly directed family of spectral subspaces. Then, the closure of $\bigcup_{i\in\mathcal{I}} M_i$ is a spectral subspace.*

Proof. Since $\{M_i\}_{i\in\mathcal{I}}$ is an increasingly directed family, $\bigcup_{i\in\mathcal{I}} M_i$ is a vector subspace. Let M be its closure in H. Then, clearly, M is doubly invariant.

Let $\varphi \to U_\varphi^i$ be the extension to $C(X)$ of the spectral representation $f \to T_f|M_i$. If M_1 and M_2 are spectral then, according to Corollary 8.3, $M_1 \cap M_2$ is also spectral and $U_\varphi^1 h = U_\varphi^2 h$ for $h \in M_1 \cap M_2$.

Let $h \in \bigcup_{i\in\mathcal{I}} M_i$ and define $U_\varphi h = U_\varphi^i h$ for $h \in M_i$. Following the above considerations, the operators U_φ are well defined on $\bigcup_{i\in\mathcal{I}} M_i$ and, for any $\varphi \in C(X)$ and $h \in \bigcup_{i\in\mathcal{I}} M_i$, we have

$$\|U_\varphi h\| \leqslant \|\varphi\|\ \|h\|.$$

Therefore, U_φ may be extended, by continuity, to M and it is easily verified that $\varphi \to U_\varphi$ is an extension to $C(X)$ of the representation $f \to T_f|M$. ◇

Theorem 8.6. *Let $f \to T_f$ be a representation of A on H. The space H has a unique decomposition of the form*

$$H = H_s \oplus H_c$$

such that H_s and H_c are doubly invariant subspaces, the representation $f \to T_f|H_s$ is spectral and the representation $f \to T_f|H_c$ is completely non-spectral. H_s is the largest spectral subspace of H.

Proof. Let $\{M_i\}_{i\in\mathcal{I}}$ be the set of all spectral subspaces of H. The subspace $\{0\}$ is, obviously, spectral, hence this set is non-void. $\{M_i\}_{i\in\mathcal{I}}$ is increasingly directed, according to Proposition 8.4. Let H_s be the closure of $\bigcup_{i\in\mathcal{I}} M_i$. Then, from Proposition 8.5, H_s is spectral and, clearly, it is the largest spectral subspace of H. We write

$$H = H_s \oplus H_c.$$

H_s is doubly invariant, hence H_c is doubly invariant too. H_s being spectral, $f \to T_f|H_s$ is also spectral and, as H_s is maximal, $f \to T_f|H_c$ is completely non-spectral.

Consider another decomposition of H

$$H = H_1 \oplus H_2$$

with H_1, H_2 doubly invariant, $f \to T_f|H_1$ spectral and $f \to T_f|H_2$ completely non-spectral. Hence, H_1 is a spectral subspace and, as H_s is maximal, $H_1 \subset H_s$.

Let

$$H_s = H_1 \oplus M.$$

Since H_s and H_1 are doubly invariant, M is also doubly invariant and, according to Corollary 8.3, M is spectral. But M is orthogonal to H_1, hence $M \subset H_2$ and, as $f \to T_f|H_2$ is completely non-spectral, there results $M = \{0\}$. Therefore

$$H_s = H_1, \quad H_c = H_2$$

and the proof is complete. ◇

The representation $f \to T_f|H_s$ will be called the spectral part of the representation $f \to T_f$, and $f \to T_f|H_c$ the completely non-spectral part.

8.2. The spectral dilation and attached spectral measures

A representation $\varphi \to U_\varphi$ of $C(X)$ on a Hilbert space K is called a *spectral dilation* of the representation $f \to T_f$ of A on H, if H is a Hilbert subspace of K and

(8.2.1) $$T_f h = PU_f h \qquad (f \in A;\ h \in H)$$

where P is the orthogonal projection of K on H.

We obviously have

(8.2.2) $$T_f^* h = PU_f^* h = PU_{\bar{f}} h \qquad (f \in A;\ h \in H)$$

The spectral dilation $\varphi \to U_\varphi$ of $f \to T_f$ is said to be *minimal* if the space generated by the elements of the form $U_\varphi h$, with $\varphi \in C(X)$ and $h \in H$, is dense in K.

If the representation $f \to T_f$ admits a spectral dilation, then it clearly also admits a minimal one. In the following all the considered spectral dilations will be assumed minimal.

Let $\varphi \to U_\varphi$ be a spectral dilation of $f \to T_f$ and $(\mu_{k_1,k_2})_{k_1,k_2} \in K$ the spectral family attached to the representation $\varphi \to U_\varphi$ of $C(X)$ on K. For $h \in H$ we put $\mu_h = \mu_{h,h}$. For any $f, g \in A$ we have

$$(T_f h, h) = (PU_f h, h) = (U_f h, h) = \int f \mathrm{d}\mu_h$$

and

$$\|(T_f + T_g^*)h\|^2 = \|PU_{f+\bar{g}}h\|^2 \leqslant \|U_{f+\bar{g}}h\|^2 = (U_{f+\bar{g}}h, U_{f+g}h) =$$

$$= (U_{|f+\bar{g}|^2}\, h, h) = \int |f + \bar{g}|^2 \,\mathrm{d}\mu_h.$$

A positive measure μ_h on X, which satisfies

$$(T_f h, h) = \int f \mathrm{d}\mu_h \qquad (f \in A) \tag{8.2.3}$$

and

$$\|(T_f + T_g^*) h\|^2 \leqslant \int |f + \bar{g}|^2 \,\mathrm{d}\mu_h \qquad (f, g \in A) \tag{8.2.4}$$

will be called a *spectral measure attached to h* by $f \to T_f$.

The measure $\mu_h = \mu_{h,h}$, $h \in H$, where $(\mu_{k_1,k_2})_{k_1,k_2 \in K}$ is the spectral family attached to a spectral dilation $\varphi \to U_\varphi$ of $f \to T_f$, is, as we have already seen, a spectral measure attached to h by $f \to T_f$.

In the following we shall study the existence of attached spectral measures and spectral dilations.

Theorem 8.7. *Let A be a Dirichlet algebra on X and $f \to T_f$ a representation of A on H. There exists a unique spectral dilation $\varphi \to U_\varphi$ of $f \to T_f$.*

Proof. From Proposition 7.8 there results

$$\|\mathrm{Re}T_f\| \leqslant \|\mathrm{Re}f\|. \tag{8.2.5}$$

Now using (8.2.5) we obtain

$$\|T_f + T_g^*\| = \|\mathrm{Re}T_f + \mathrm{i}\mathrm{Im}T_f + \mathrm{Re}T_g - \mathrm{i}\mathrm{Im}T_g\| = \|\mathrm{Re}T_{f+g} +$$

$$+ \mathrm{i}\,(\mathrm{Re}T_{-\mathrm{i}f+\mathrm{i}g})\| \leqslant \|\mathrm{Re}T_{f+g}\| + \|\mathrm{Re}T_{-\mathrm{i}f+\mathrm{i}g}\| \leqslant \|\mathrm{Re}(f+g)\| +$$

$$+ \|\mathrm{Re}\,(-\mathrm{i}f + \mathrm{i}g)\| \leqslant \|\mathrm{Re}\,(f + \bar{g})\| + \|\mathrm{Re}\,(-\mathrm{i}f + \mathrm{i}g)\| \leqslant 2\|f + \bar{g}\|$$

which yields

$$\|T_f + T_g^*\| \leqslant 2\|f + \bar{g}\|. \tag{8.2.6}$$

For $h_1, h_2 \in H$ we define on $A + \bar{A}$ the functional

$$\mu_{h_1, h_2}(f + \bar{g}) = ((T_f + T_g^*)\, h_1, h_2).$$

According to (8.2.6), μ_{h_1,h_2} is well defined on $A + \bar{A}$ and bounded; therefore, since A is a Dirichlet algebra on X, μ_{h_1,h_2} may be extended to some Radon measure μ_{h_1,h_2} on X. One easily verifies that the family $(\mu_{h_1,h_2})_{h_1,h_2 \in H}$ satisfies (7.1.2), hence it is a semi-spectral family.

Applying Theorem 7.5, we obtain a representation $\varphi \to U_\varphi$ of $C(X)$ on a Hilbert space K which is a spectral dilation of $f \to T_f$.

Let $\varphi \to U_\varphi$ and $\varphi \to U'_\varphi$ be two spectral dilations of $f \to T_f$ and $(\mu_{k_1,k_2})_{k_1,k_2 \in K}$, $(\mu'_{k'_1,k'_2})_{k'_1,k'_2 \in K'}$ be the corresponding spectral families. For $f, g \in A$ and $h_1 h_2 \in H$, we have

$$\mu_{h_1,h_2}(f + \bar{g}) = (U_{f+\bar{g}}h_1, h_2) = (U_f h_1, h_2) + (U_g^* h_1, h_2) =$$

$$= (PU_f h_1, h_2) + (PU_g^* h_1, h_2) = (T_f h_1, h_2) + (T_g^* h_1, h_2) =$$

$$= (PU'_f h_1, h_2) + (PU'_{\bar{g}} h'_1, h_2) = (U'_{f+\bar{g}} h_1, h_2) =$$

$$= \mu'_{h_1,h_2}(f \dot{+} \bar{g}).$$

Hence, the measures μ_{h_1,h_2} and μ'_{h_1,h_2} are equal on $A + \bar{A}$ and, since A is a Dirichlet algebra, they are equal. Since the spectral families

$(\mu_{k_1, k_2})_{k_1, k_2 \in K}$ and $(\mu'_{k'_1, k'_2})_{k'_1, k'_2 \in K'}$ are minimal dilations of the semi-spectral families $(\mu_{h_1, h_2})_{h_1, h_2 \in H}$, $(\mu'_{h_1, h_2})_{h_1, h_2 \in H}$, respectively, then according to Theorem 7.6, they are equal (in the sense precised in Theorem 7.6). This proves the uniqueness of the spectral dilation in the case of a Dirichlet algebra.

The theorem is completely proved. ◇

According to formula (8.2.5), which holds for any representation $f \to T_f$ on H of a function algebra A, if for any $\varphi \in A + \bar{A}$ of the form $\varphi = f + \bar{g}, f, g \in A$, we put

$$T_\varphi = T_f + T_g^*,$$

then the map $\varphi \to T_\varphi$ of $A + \bar{A}$ in $L(H)$ is well defined, linear and bounded.

We now prove the existence of the spectral measure attached to h, for the representations $f \to T_f$ of A which satisfy the condition

$$\sum \|T_{\varphi_i} h\|^2 \leqslant \|h\|^2, \tag{8.2.7}$$

for any finite system $\varphi_1, \ldots, \varphi_n$ of elements in $A + \bar{A}$, with $\Sigma|\varphi_i|^2 \leqslant 1$.

Proposition 8.8. *Let $f \to T_f$ be a representation of A on H such that (8.2.7) holds for $h \in H$. Then there exists a positive measure μ_h on X such that $\mu_h(1) = \|h\|^2$ and*

$$\|(T_f + T_g^*)h\|^2 \leqslant \int |f + \bar{g}|^2 \, d\mu_h. \tag{8.2.8}$$

Proof. For $g \in C(X)$, $g \geqslant 0$, we put

$$\mu_*(g) = \sup_{\Sigma|\varphi_i|^2 \leqslant g} \sum \|T_{\varphi_i} h\|^2 \tag{8.2.9}$$

where the supremum is taken for all finite systems $\varphi_1, \ldots, \varphi_n$ of elements in $A + \bar{A}$, with $\Sigma|\varphi_i|^2 \leqslant g$.

We have

$$\mu_*(\alpha g) = \alpha \mu_*(g) \quad \text{for} \quad \alpha \geqslant 0 \tag{8.2.10}$$

and from (8.2.7) there results

(8.2.11) $$\mu_*(g) \leqslant \|g\|\, \|h\|^2.$$

For $g_1, g_2 \in C_R(X)$, $g_1, g_2 \geqslant 0$ we have

$$\mu_*(g_1 + g_2) = \sup_{\Sigma|\varphi_i|^2 \leqslant g_1+g_2} \sum \|T_{\varphi_i} h\|^2 \geqslant \sup_{\Sigma|\varphi_i|^2 \leqslant g_1} \sum \|T_{\varphi_i} h\|^2 +$$

$$+ \sup_{\Sigma|\varphi_i|^2 \leqslant g_2} \sum \|T_{\varphi_i} h\|^2 = \mu_*(g_1) + \mu_*(g_2)$$

that is

(8.2.12) $$\mu_*(g_1 + g_2) \geqslant \mu_*(g_1) + \mu_*(g_2).$$

Let us note that

(8.2.13) $$\mu_*(1) = \|h\|^2.$$

For some $g \in C_R(X)$ we write

(8.2.14) $$\mu^*(g) = \inf_{c > |g|} (c\|h\|^2 - \mu_*(c - |g|)).$$

From (8.2.12) and (8.2.13) we find that

$$\mu_*(|g|) + \mu_*(c - |g|) \leqslant \mu_*(c) = c\|h\|^2$$

for any $c > |g|$. Hence

$$c\|h\|^2 - \mu_*(c - |g|) \geqslant \mu_*(|g|) \geqslant 0$$

for any $c > |g|$. Therefore, for any $g \in C_R(X)$,

(8.2.15) $$\mu^*(g) \geqslant 0.$$

It is clear that for $\alpha > 0$, we have

$$\mu^*(\alpha g) = \alpha\mu^*(g). \tag{8.2.16}$$

Let now $g_1, g_2 \in C_R(X)$ and c_1, c_2 be two constants, with $c_1 > |g_1|$, $c_2 > |g_2|$, and

$$c_1\|h\|^2 - \mu_*(c_1 - |g_1|) \leqslant \mu^*(g_1) + \varepsilon$$

$$c_2\|h\|^2 - \mu_*(c_2 - |g_2|) \leqslant \mu^*(g_2) + \varepsilon.$$

Then $c_1 + c_2 \geqslant |g_1 + g_2|$ and

$$(c_1 + c_2)\,\|h\|^2 - \mu_*(c_1 + c_2 - |g_1 + g_2|) \leqslant$$

$$\leqslant (c_1 + c_2)\,\|h\|^2 - \mu_*(c_1 + c_2 - |g_1| - |g_2|) \leqslant$$

$$\leqslant c_1\|h\|^2 - \mu_*(c_1 - |g_1|) + c_2\|h\|^2 - \mu_*(c_2 - |g_2|) \leqslant$$

$$\leqslant \mu^*(g_1) + \mu^*(g_2) + 2\varepsilon.$$

We used the fact that μ^* is a positive functional, and (8.2.12). Hence

$$\mu^*\,(g_1 + g_2) \leqslant \mu^*(g_1) + \mu^*(g_2) \tag{8.2.17}$$

for any $g_1, g_2 \in C_R(X)$.

From (8.2.14) there follows

$$\mu^*(g) \leqslant \|g\|\,\|h\|^2 - \mu^*(\|g\| - |g|) \leqslant \|g\|\,\|h\|^2 \tag{8.2.18}$$

and therefore

$$\mu^*(1) = \|h\|^2. \tag{8.2.19}$$

From (8.2.15), (8.2.16), (8.2.17) and (8.2.18), μ^* is a seminorm on $C_R(X)$ and, according to Hahn-Banach theorem and (8.2.19), there exists a real measure μ_h on X such that $\mu_h(1) = \|h\|^2$ and

$$|\mu_h(g)| \leqslant |\mu^*(g)| \leqslant \|g\| \, \|h\|^2. \tag{8.2.20}$$

Then, obviously

$$\|\mu_h\| = \mu_h(1) = \|h\|^2$$

hence μ_h is a positive measure on X.
We have

$$\mu_h(g) \leqslant \mu^*(g) \leqslant \|h\|^2 c - \mu_*(c - g)$$

for $g \geqslant 0$ and $c \geqslant g$, and therefore

$$\mu_h(c - g) \geqslant \mu_*(c - g)$$

for any $g \geqslant 0$ and $c \geqslant g$. Since any positive function in $C_R(X)$ may be written under the form $c - g$ with $g \geqslant 0$, $c \geqslant g$, then for any $g \in C_R(X)$, $g \geqslant 0$, we have

$$\mu_h(g) \geqslant \mu_*(g). \tag{8.2.21}$$

Then

$$\mu_h(f + \bar{g})^2 \geqslant \mu_*(f + \bar{g}) \geqslant \|T_f h + T_g^* h\|^2$$

that is

$$\|T_f h + T_g^* h\|^2 \leqslant \int |f + \bar{g}|^2 d\mu_h$$

and the proposition is completely proved. ◇

Proposition 8.9. *Let μ_h be a positive measure on X such that $\mu_h(1) = \|h\|^2$ and*

$$\|T_f h + T_g^* h\|^2 \leqslant \int |f + \bar{g}|^2 d\mu_h \qquad (f, g \in A). \tag{8.2.22}$$

Then μ_h is a spectral measure attached to h by $f \to T_f$.

Proof. It remains to prove only that

$$(T_f h, h) = \int f d\mu_h. \tag{8.2.23}$$

Let λ be an arbitrary complex number and $f \in A$. From (8.2.22) there results

$$\|T_f h + \bar{\lambda} h\|^2 \leqslant \int |f + \bar{\lambda}|^2 d\mu_f,$$

hence

$$\|T_f h\|^2 + 2\mathrm{Re}\lambda(T_f h, h) + |\lambda|^2 \|h\|^2 \leqslant \int |f|^2 d\mu_h + 2\mathrm{Re}[\lambda \int f d\mu_h] +$$
$$+ |\lambda|^2 \|h\|^2$$

or

$$2\mathrm{Re}[\lambda((T_f h, h) - \int f d\mu_h)] \leqslant \int |f|^2 d\mu_k - \|T_f h\|^2. \tag{8.2.24}$$

Since (8.2.24) holds for any complex number λ, we get

$$(T_f h, h) = \int f d\mu_h$$

and the proof is complete. ◇

From Proposition 8.8 and 8.9 there results the following

Theorem 8.10. *Let $f \to T_f$ be a representation of the algebra A on H and $h \in H$ for which* (8.2.7) *holds. Then there exists a spectral measure attached to h by $f \to T_f$.*

In the case of logmodular algebras we shall prove the uniqueness of attached spectral measures together with the existence and uniqueness of the spectral dilation when there are spectral measures attached to every point h of H.

Theorem 8.11. *If A is logmodular on X and $f \to T_f$ is a representation of A on H, then the spectral measure attached to $h \in H$ by $f \to T_f$ is unique.*

Proof. Let μ_1, μ_2 be two spectral measures attached to $h \in H$ by $f \to T_f$. For $f, f^{-1} \in A$ we have

$$\|h\|^4 = (h, h)^2 = (T_{f^{-1}}T_f h, h)^2 = (T_f h, T^*_{f^{-1}}h)^2 \leqslant$$

$$\leqslant \|T_f h\|^2 \, \|T^*_{f^{-1}}h\|^2 \leqslant \int |f|^2 \mathrm{d}\mu_1 \int |f^{-1}|^2 \, \mathrm{d}\mu_2.$$

Since A is a logmodular algebra, we have

$$\|h\|^4 \leqslant \int e^{2u}\mathrm{d}\mu_1 \int e^{-2u}\mathrm{d}\mu_2$$

for any $u \in C_R(X)$. Hence, the 'eal function

$$\varphi(t) = \int e^{2tu}\mathrm{d}\mu_1 \int e^{-2tu}\mathrm{d}\mu_2$$

has an extreme point in 0 for any $u \in C_R(X)$ and therefore

$$0 = \varphi'(0) = 2\left(\int u\mathrm{d}\,\mu_1 - \int u\mathrm{d}\,\mu_2\right)$$

for any $u \in C_R(X)$, i.e. $\mu_1 = \mu_2$. ◇

Theorem 8.12. *Let A be a logmodular algebra on X and $f \to T_f$ a representation of A on H such that for any $h \in H$ there exists a spectral measure attached to h by $f \to T_f$. Then there exists a unique spectral dilation of $f \to T_f$.*

Proof. Let μ_h be the spectral measure attached to h by $f \to T_f$. We first prove the following relation

$$\mu_{h_1+h_2} + \mu_{h_1-h_2} = 2\mu_{h_1} + 2\mu_{h_2} \qquad (h_1, h_2 \in H). \tag{8.2.25}$$

Let $f, f^{-1} \in A$. We have

$$2(\|h_1\|^2 + \|h_2\|^2) = \|h_1 + h_2\|^2 + \|h_1 - h_2\|^2 =$$

$$= (T_f(h_1 + h_2), T^*_{f^{-1}}(h_1 + h_2)) + (T_f(h_1 - h_2), T^*_{f^{-1}}(h_1 - h_2)) \leqslant$$

$$\leqslant \|T_f(h_1 + h_2)\| \ \|T^*_{f^{-1}}(h_1 + h_2)\| + \|T^*_{f^{-1}}(h_1 - h_2)\| \ \|T_f(h_1 - h_2)\| \leqslant$$

$$\leqslant [\|T_f(h_1 + h_2)\|^2 + \|T_f(h_1 - h_2)\|^2]^{1/2}[\|T^*_{f^{-1}}(h_1 + h_2)\|^2 +$$

$$+ \|T^*_{f^{-1}}(h_1 - h_2)\|^2]^{1/2} = [\|T_f(h_1 + h_2)\|^2 + \|T_f(h_1 - h_2)\|^2]^{1/2}.$$

$$\cdot [2\|T^*_{f^{-1}}h_1\|^2 + 2\|T^*_{f^{-1}}h_2\|^2]^{1/2} \leqslant$$

$$\leqslant [\textstyle\int |f|^2 \mathrm{d}(\mu_{h_1+h_2} + \mu_{h_1-h_2})]^{1/2} \left[\displaystyle\int \frac{1}{|f|^2} \mathrm{d}(2\mu_{h_1} + 2\mu_{h_2})\right]^{1/2}.$$

Since A is a logmodular algebra we get

$$4(\|h_1\|^2 + \|h_2\|^2)^2 \leqslant [\textstyle\int e^{2u} \mathrm{d}(\mu_{h_1+h_2} + \mu_{h_1-h_2})] \times$$

$$\times [\textstyle\int e^{-2u} \mathrm{d}\,(2\mu_{h_1} + 2\mu_{h_2})]$$

for any $u \in C_R(X)$. Then, the standard argument used also in the proof of the preceding theorem, yields (8.2.25).

If we put

$$4\mu_{h_1, h_2} = \mu_{h_1+h_2} - \mu_{h_1-h_2} + i\mu_{h_1+ih_2} - i\mu_{h_1-ih_2}$$

we get a semispectral family $(\mu_{h_1, h_2})_{h_1, h_2 \in H}$ on X such that

$$(T_f h_1, h_2) = \textstyle\int f \mathrm{d}\mu_{h_1, h_2}.$$

Using Theorem 7.5 we obtain a representation $\varphi \to U_\varphi$ of $C(X)$ on a Hilbert space K, which is a spectral dilation of $f \to T_f$.

Let $\varphi \to U'_\varphi$ be another spectral dilation of $f \to T_f$. Writing

$$m_{h_1,h_2}(\varphi) = (U'_\varphi h_1, h_2)$$

for $h_1, h_2 \in H$, then, as we have seen, the measure $m_h = m_{h,h}$ is a spectral measure attached to h by $f \to T_f$. From Theorem 8.11 there follows $m_h = \mu_h$ hence, the semispectral families $(m_{h_1,h_2})_{h_1,h_2 \in H}$ and $(\mu_{h_1,h_2})_{h_1,h_2 \in H}$ coincide.

Since the spectral dilation for semispectral families is unique (Theorem 7.6), the spectral dilations $\varphi \to U_\varphi$ and $\varphi \to U'_\varphi$ of $f \to T_f$ coincide (in the sense given above).

The theorem is proved. ◇

Theorem 8.13. *Let A be a logmodular algebra on X and $f \to T_f$ a representation of A on H. The representation $f \to T_f$ has a spectral dilation if and only if (8.2.7) holds for any $h \in H$. If $f \to T_f$ has a spectral dilation then it is unique.*

Proof. If (8.2.7) holds for any $h \in H$ then, from Theorem 8.10 there results that $f \to T_f$ has a spectral dilation. The uniqueness of this spectral dilation follows from Theorem 8.12.

Let $\varphi \to U_\varphi$ be a spectral dilation of $f \to T_f$ and $\varphi_1, \ldots, \varphi_n$ a finite system of elements in $A + \bar{A}$, of the form $\varphi = f_\iota + \bar{g}_i$, with $\Sigma |\varphi_i|^2 \leqslant 1$ and $h \in H$. We have

$$\sum \|T_{\varphi_\iota} h\|^2 = \sum \|(T_{f_\iota} + T^*_{g_\iota}) h\|^2 = \sum \|P(U_{f_\iota + \bar{g}_\iota}) h\|^2 \leqslant$$

$$\leqslant \sum \|U_{\varphi_\iota} h\|^2 \leqslant \sum \|\varphi_i\|^2 \|h\|^2 \leqslant \|h\|^2.$$

The proof is therefore complete. ◇

The condition (8.2.7) in Theorem 8.13 can be reduced to a much simpler one. Indeed we have the following.

Proposition 8.14. *Let A be a logmodular algebra on X and $f \to T_f$ a representation of A on H. Then the following two conditions are equivalent.*

(i) *$f \to T_f$ satisfies (8.2.7) for all $h \in H$.*

(ii) *For all $h \in H$ and $f, g \in A$ we have*

$$|f|^2 + |g|^2 \leqslant 1 \Rightarrow \begin{cases} \|T_f h\|^2 + \|T_g h\|^2 \leqslant \|h\|^2 \\ \|T^*_f h\|^2 + \|T^*_g h\|^2 \leqslant \|h\|^2 \end{cases} \tag{8.2.26}$$

Proof. The implication (i) ⇒ (ii) is obvious. So we can suppose that condition (ii) holds. Firstly, we shall prove (by induction) that

(8.2.27) $$\sum_{i=1}^{n} |f_i|^2 \leqslant 1 \Rightarrow \sum_{i=1}^{n} \|T_{f_i}h\|^2 \leqslant \|h\|^2$$

for any finite system $\{f_1,\ldots,f_n\}$ of elements in A and $h \in H$. For $n = 2$ (8.2.27) is contained in (8.2.26). Let thus $n > 2$, $\varepsilon > 0$ and $f_1,\ldots,f_n \in A$ such that $\sum_{i=1}^{n} |f_i|^2 < 1$. Since A is logmodular algebra on X, we can find $g \in A$ such that $g^{-1} \in A$ and

(8.2.28) $$\sum_{i=1}^{n-1} |f_i|^2 < |g|^2 < 1 - |f_n|^2 + \varepsilon$$

Let $g_i = f_i g^{-1}$, i = 1, 2,..., $n - 1$. Then, by (8.2.28) $\sum_{i=1}^{n-1} |g_i|^2 \leqslant 1$ thus by the induction hypothesis

$$\sum_{i=1}^{n-1} \|T_{g_i}k\|^2 \leqslant \|k\|^2$$

for all $k \in H$. Particularly, for $k = T_g h$ we obtain

$$\sum_{i=1}^{n-1} \|T_{f_i}h\|^2 \leqslant \|T_g h\|^2$$

But, from (8.2.28) we have also

$$|f_n|^2 + |g|^2 \leqslant 1 + \varepsilon$$

so that using (8.2.26) we infer

$$\|T_{f_n}h\|^2 + \|T_g h\|^2 \leqslant (1 + \varepsilon) \|h\|^2$$

that is

$$\sum_{i=1}^{n} \|T_{f_i}h\|^2 \leqslant (1 + \varepsilon) \|h\|^2$$

Letting $\varepsilon \to 0$, we obtain finally (8.2.27).

In a similar way we infer that

$$\sum_{i=1}^{n} |f_i|^2 \leqslant 1 \Rightarrow \sum_{i=1}^{n} \|T^*_{f_i}h\|^2 \leqslant \|h\|^2 \tag{8.2.29}$$

for any finite system $\{f_1,\ldots,f_n\}$ in A and $h \in H$.

Now define for $g \in C_R(X)$, $g \geqslant 0$

$$\mu_*(g) = \sup \sum \|T_{f_i}h\|^2$$

where the supremum is taken for all finite systems $\{f_1,\ldots,f_n\} \subset A$ such that $\Sigma|f_i|^2 \leqslant g$. Then reproducing the proof of Proposition 8.8 we obtain (using (8.2.27) instead of (8.2.7)) a positive measure μ_1 on X such that $\mu_1(1) = \|h\|^2$ and

$$\|T_f h\|^2 \leqslant \int |f|^2 \mathrm{d}\mu_1, \qquad (f \in A) \tag{8.2.30}$$

Analogously, (8.2.29) will lead (again by reproducing the proof of Proposition 8.8) to the existence of a positive measure μ_2 on X such that $\mu_2(1) = \|h\|^2$ and

$$\|T^*_f h\|^2 \leqslant \int |f|^2 \mathrm{d}\mu_2, \qquad (f \in A) \tag{8.2.31}$$

In virtue of the proof of the Proposition 8.9 we will have

$$(T_f, h, h) = \int f \mathrm{d}\mu_1 = \int f \mathrm{d}\mu_2, \qquad (f \in A) \tag{8.2.32}$$

Actually $\mu_1 = \mu_2$ since for $f, f^{-1} \in A$ we have

$$\|h\|^4 \leqslant (T_f h, T^*_{f^{-1}} h)^2 \leqslant \|T_f h\|^2 \|T^*_{f^{-1}} h\|^2 \leqslant$$

$$\leqslant \int |f|^2 \mathrm{d}\mu_1 \int |f|^{-2} \mathrm{d}\mu_2.$$

The fact that $\mu_1 = \mu_2$ is easily obtained as in the proof of Theorem 8.11. Let us put $\mu_h = \mu_1 = \mu_2$.

Finally if $\varphi_i = f_i + \bar{g}_i$, with $\Sigma|\varphi_i|^2 \leqslant 1$ and $f_1,\ldots,f_n, g_1,\ldots, g_n \in A$ then

$$\sum_{i=1}^{n} \|T_{\varphi_i}h\|^2 = \sum_{i=1}^{n} \|T_{f_i}h + T^*_{g_i}h\|^2 \leqslant \sum_{i=1}^{n} (\|T_{f_i}h\|^2 +$$

$$+ 2\mathrm{Re}\,(T_{f_i g_i}h, h) + \|T_{g_i}h\|^2) = \sum_{i=1}^{n} (\|T_{f_i}h\|^2 +$$

$$+ 2\mathrm{Re} \int f_i g_i \mathrm{d}\mu_h + \|T^*_{g_i} h\|^2) \leqslant \sum_{i=1}^{n} (\int |f_i|^2 \mathrm{d}\mu_h +$$

$$+ 2\mathrm{Re} \int f_i g_i \mathrm{d}\mu_h + \int |g_i|^2 \mathrm{d}\mu_h) = \int \left(\sum_{i=1}^{n} |\varphi_i|^2 \ \mathrm{d}\mu_h \right) \leqslant \|h\|^2.$$

The proof of the proposition is complete. ◇

8.3. Szegö measures and natural representations

Let μ be a positive measure on X. We recall that $H^2(\mathrm{d}\mu)$ is the Hilbert space obtained by closing A in $L^2(\mathrm{d}\mu)$. The measure μ is called a *Szegö measure* (relative to A) if, for any Borel set E of X for which

$$\chi_E L^2(\mathrm{d}\mu) \subset H^2(\mathrm{d}\mu),$$

we have $\mu(E) = 0$ ($\chi_E =$ the characteristic function of E).

The following theorem asserts the existence of Szegö measure for any algebra $A \neq C(X)$:

Theorem 8.15. (i) *Any representing measure of A is either a Szegö measure or a point measure.*

(ii) *if ν is a non-zero complex measure on X, orthogonal to A, then $|\nu|$ is a Szegö measure.*

(iii) *If $A \neq C(X)$ then there exist Szegö measures on X (relative to A).*

Proof. (i) Let μ be a representing measure of A. One easily verifies that μ is multiplicative on $H^2(\mathrm{d}\mu)$. Assume that for a Borel set E we have

$$\chi_E L^2(\mathrm{d}\mu) \subset H^2(\mathrm{d}\mu). \tag{8.3.1}$$

Then $\chi_E \in H^2(\mathrm{d}\mu)$ and

$$\int \chi_E \mathrm{d}\mu = \int \chi_E^2 \mathrm{d}\mu = [\int \chi_E \mathrm{d}\mu]^2$$

Hence $\mu(E) = 0$ or $\mu(E) = 1$. If $\mu(E) = 1$ then from (8.3.1), there results $L^2(\mathrm{d}\mu) = H^2(\mathrm{d}\mu)$, hence μ is multiplicative on $L^2(\mathrm{d}\mu)$ and, in particular, on $C(X)$. It is known that such measures are point measures.

(ii) Let ν be a complex measure, orthogonal to A. Then ν is orthogonal to $H^2(\mathrm{d}\mu)$ and, writing $\mu = |\nu|$, we have

$$\mathrm{d}\nu = F\mathrm{d}\mu, \qquad F \in L^\infty(\mathrm{d}\mu), \qquad |F| = 1 \tag{8.3.2}$$

Suppose $\chi_E L^2(d\mu) \subset H^2(d\mu)$. There results $\chi_E \bar{F} \in H^2(d\mu)$ and following (8.3.2), we have

$$\mu(E) = \int \chi_E d\mu = \int \chi_E F\bar{F} d\mu = \int \chi_E \bar{F} d\nu = 0.$$

(iii) is a consequence of (ii) and of the Hahn-Banach theorem. The proof is therefore complete. ◇

A positive measure μ on X be called *Szegö – singular* (relative to A) if $L^2(d\mu) = H^2(d\mu)$.

We shall now establish the relation between the Szegö measure and a completely nonspectral representation.

Theorem 8.16. *Let $f \to T_f$ be a representation of A on H and μ_h a spectral measure attached to the element $h \in H$. Let M be a closed subspace of H such that $h \in M$ and $T_f M \subset M$ for any $f \in A$. If the representation $f \to T_f|M$ of A on M is completely non-spectral, then μ_h is a Szegö measure.*

Proof. Let $V_f = T_f|M$; we define the operator V on $A + \bar{A}$, with values in M, by

$$V(f + \bar{g}) = V_f h + V_g^* h \qquad (f, g \in A).$$

Since μ_h is a spectral measure attached to h by $f \to T_f$, we have

$$\|V(f + \bar{g})\|^2 = \|V_f h + V_g^* h\|^2 \leqslant \|T_f h + T_g^* h\|^2 \leqslant$$

$$\leqslant \int |f + \bar{g}|^2 d\mu_h.$$

Hence V may be extended to a contraction (denoted also by V) from the space K, the closure of the set spanned by $A + \bar{A}$ in $L^2(d\mu)$, into M.

For $\varphi \in C(X)$ we denote by U_φ the multiplication by φ in $L^2(d\mu)$. Then, for $f, g \in A$,

$$V(U_f g) = V(fg) = V_{fg} h = V_f V_g h = V_f V g \quad (f, g \in A)$$

and, by continuity

$$(8.3.3) \qquad V(U_f | H^2(d\mu_h)) = V_f(V | H^2(d\mu_h)).$$

Similarly, we find

$$V(U_f^* | \overline{H^2(d\mu_h)}) = V_f^*(V | \overline{H^2(d\mu_h)}), \tag{8.3.4}$$

where

$$\overline{H^2(d\mu_h)} = \{\bar{h}: h \in H^2(d\mu_h)\}$$

Let E be a Borel set of X, such that

$$\chi_E L^2(d\mu_h) \subset H^2(d\mu_h).$$

Then obviously,

$$\chi_E L^2(d\mu_h) \subset H^2(d\mu_h) \cap \overline{H^2(d\mu_h)},$$

as $\chi_E L^2(d\mu_h)$ is doubly invariant to $\varphi \to U_\varphi$, then, according to (8.3.3) and (8.3.4), the M-closure of $V\chi_E L^2(d\mu_h)$ is a doubly invariant subspace N and

$$V(U_f | \chi_E L^2(d\mu_h)) = V_f(V | \chi_E L^2(d\mu_h)) \tag{8.3.5}$$

$$V(U_f^* | \chi_E L^2(d\mu_h)) = V_f^*(V | \chi_E L^2(d\mu_h)).$$

Following (8.3.5), we have

$$V_g^* V_f V e = V_g^* V U_f e = V U_g^* U_f e = V U_{\bar{g}f} e = V U_{f\bar{g}} e =$$

$$= V U_f U_g^* e = V_g V_g^* V e$$

for any $e \in X_E L^2(d\mu_h)$, and by continuity

$$V_g^* V_f n = V_f V_g^* n \qquad (f, g \in A;\ n \in N). \tag{8.3.6}$$

For $\varphi \in C(X)$ of the form

$$\varphi = \sum \bar{f}_i g_i \tag{8.3.7}$$

with f_i, $g_i \in A$, and $n \in N$, we put

$$V'_\varphi n = \sum V_{f_i}^* V_{g_i} n. \tag{8.3.8}$$

Then, using (8.3.5), we obtain

$$V'_\varphi Ve = \sum V^*_{f_i} V_{g_i} Ve = V(\varphi e) \tag{8.3.9}$$

for $e \in \chi_E L^2(\mathrm{d}\mu_h)$.

Hence, if $\varphi = 0$ then $V'_\varphi Ve = 0$ for any $e \in \chi_E L^2(\mathrm{d}\mu_h)$ and, by continuity, $V'_\varphi = 0$. Therefore $\varphi \to V'_\varphi$ is a well-defined linear map on the elements φ of the form (8.3.7). Also from (8.3.9) we get

$$V'_{\varphi_1} V'_{\varphi_2} = V'_{\varphi_1 \varphi_2} \tag{8.3.10}$$

and, following (8.3.9), (8.3.10), there results

$$\| V'^m_\varphi Ve \| = \| V'^m_\varphi Ve \| = \| V(\varphi^m e) \| \leqslant$$
$$\leqslant \int |\varphi|^2 |e|^{2m} \mathrm{d}\mu_h = \|\varphi\|^m \|e\|$$

where $\|\varphi\|$ resp. $\|e\|$ are the norms in $C(X)$ resp. $L^2(\mathrm{d}\mu_h)$. Therefore

$$\| V'^m_\varphi Ve \| \leqslant \|\varphi\|^m \|e\|. \tag{8.3.11}$$

Since V'_φ are normal operators, we have

$$\| V'_\varphi Ve \| = (V'_\varphi Ve, V'_\varphi Ve)^{1/2} = (V'^*_\varphi V'_\varphi Ve, Ve)^{1/2} \leqslant$$
$$\leqslant \| V'^*_\varphi V'_\varphi Ve \|^{1/2} \| Ve \|^{1/2} = (V'_\varphi V'_\varphi Ve, V'^*_\varphi V'_\varphi Ve)^{1/2^2} \| Ve \|^{1/2} =$$
$$= (V^{*\prime 2}_\varphi V'^2_\varphi Ve, Ve)^{1/2^2} \| Ve \|^{1/2} \leqslant$$
$$\leqslant \| V'^{*2}_\varphi V'^2_\varphi Ve \| \; \| Ve \|^{\frac{1}{2} + \frac{1}{2^2}}.$$

Taking into account (8.3.11) we obtain

$$\| V'_\varphi Ve \| \leqslant \| Ve \|^{\frac{1}{2} + \frac{1}{2^2} + \cdots + \frac{1}{2^m}} \|\varphi\| \; \|e\|^{\frac{1}{2^m}} \tag{8.3.12}$$

and, for $m \to \infty$, therefore results

$$\| V'_\varphi Ve \| \leqslant \| Ve \| \; \|\varphi\|$$

hence

$$\| V'_\varphi \| \leqslant \|\varphi\|. \tag{8.3.13}$$

Since the set of all functions of the form (8.3.7) is dense in $C(X)$ then, according to (8.3.13), we may extend the map $\varphi \to V'_\varphi$ to a representation of $C(X)$ on N. As $V_f = V'_f$ for $f \in A$, the subspace N results spectral for the representation $f \to V_f$.

But $f \to V_f$ is a completely nonspectral representation, hence $N = \{0\}$. Therefore $Ve = 0$ for any $e \in \chi_E L^2(\mathrm{d}\mu_h)$. Since $\chi_E L^2(\mathrm{d}\mu_h) \subset \subset H^2(\mathrm{d}\mu_h)$, for any $e \in \chi_E L^2(\mathrm{d}\mu_h)$ there exists a sequence f_n of functions in A, such that $e = \lim_{L^2} f_n$.

We have

$$0 = (Ve, h) = \lim (Vf_n, h) = \lim (V_{f_n}h, h) =$$

$$= \lim (T_{f_n}h, h) = \lim \int f_n \mathrm{d}\mu_h = \int e \mathrm{d}\mu_h$$

for any $e \in \chi_E L^2(\mathrm{d}\mu_h)$. In particular

$$0 = \int \chi_E \mathrm{d}\mu_h = \mu_h(E).$$

Therefore μ_h is a Szegö measure. The theorem is proved. ◇

In the following we shall study a class of representations of A on H for which the converse assertion is also true.

The representation $f \to T_f$ of A on H will be called *subspectral* if there exists a spectral dilation $\varphi \to U_\varphi$ of $f \to T_f$ such that $U_f h \in H$ for any $f \in A$ and $h \in H$.

We obviously have

(8.3.14) $$T_f h = U_f h \qquad (f \in A;\ h \in H)$$

One easily verifies that any other spectral dilation of the subspectral representation $f \to T_f$, which satisfies (8.3.14), coincides with $\varphi \to U_\varphi$.

In the following by spectral dilation of the subspectral representation $f \to T_f$, we mean the (unique) spectral dilation $\varphi \to U_\varphi$ of $f \to T_f$ which satisfies (8.3.14).

Proposition 8.17. *Let $f \to T_f$ be a subspectral representation of A on H and $M \subset H$ a doubly invariant subspace to $f \to T_f$. The representation $f \to T_f | M$ of A on M is subspectral.*

Proof. Let $\varphi \to U_\varphi$ be the spectral dilation of $f \to T_f$ and K the dilation space. If we put

$$K(M) = \mathrm{clm}\, \{U_\varphi m;\ m \in M,\ \varphi \in C(X)\}$$

then $\varphi \to U_\varphi | K(M)$ is clearly a spectral dilation of $f \to T_f | M$, which satisfies (8.3.14) for $f \in A$ and $m \in M$. ◇

An important class of subspectral representations is the class of *natural representations.*

Let μ be a positive measure on X, $\mu(X) = 1$ and N a Hilbert space.

We have already seen in paragraph 7.3 how the natural representation $\varphi \to U_\varphi$ of $C(X)$ on $L^2(N; d\mu)$ is constructed:

$$U_\varphi k = \varphi k \qquad (\varphi \in C(X);\ k \in L^2(N; d\mu))$$

If we consider the elements of N as functions with constant values from X to N, we can consider N as a subspace of $L^2(N; d\mu)$.

Let

$$H^2(N; d\mu) = H^2(N; A; d\mu) = \text{clm}\,[\{U_f n, f \in A, n \in N\}].$$

It is clear that, for any $f \in A$, we have

$$U_f(H^2(N; d\mu)) \subset H^2(N; d\mu).$$

If we write $T_f = U_f | H^2(N; d\mu)$, one easily verifies that $f \to T_f$ is a subspectral representation of A on $H^2(N; d\mu)$, $\varphi \to U_\varphi$ being its spectral dilation. We call this representation the *natural representation* of A on $H^2(N; d\mu)$. If N is one-dimensional we shall write $H^2(d\mu)$ instead of $H^2(N; d\mu)$, and it is easy to see that this notation is in agreement with that of the previous chapters.

Now let $f \to T_f$ be a subspectral representation of A on H and $\varphi \to U_\varphi$ its spectral dilation.

Let $(\mu_{k_1, k_2})_{k_1, k_2 \in K}$ be the spectral family attached to the representation $\varphi \to U_\varphi$ of $C(X)$ on K(K is the dilation space) and $\mu_h = \mu_{h,h}$ the spectral measure attached to h by $f \to T_f$, for any $h \in H$. We have

$$\|T_f h\|^2 = (T_f h, T_f h) = (U_f h, U_f h) = (U_{|f|^2} h, h) = \int |f|^2 \, d\mu_h$$

that is

$$\|T_f h\|^2 = \int |f|^2 d\mu_h. \tag{8.3.15}$$

For $h \in H$, we denote by M_h the closed subspace of H spanned by $T_f h, f \in A$. Using (8.3.15) we may uniquely define the operator V on M_h, with values in $L^2(d\mu_h)$, such that

$$V(T_f h) = f.$$

The operator V is an isometry from M_h onto $V(M_h)$. It is clear that $V(M_h)$ is invariant to the multiplication by functions of A and

$$T_f m = V^{-1} f V m \qquad (f \in A;\ m \in M_h). \tag{8.3.16}$$

Proposition 8.18. *The closed subspace M of M_h is spectral with respect to $f \to T_f|M_h$ if and only if there exists a Borel subset E of X such that $VM = \chi_E L^2(\mathrm{d}\mu_h)$. E is μ_h-essentially unique determined by M.*

Proof. Let $V_f = T_f|M_h$. Assume $VM = \chi_E L^2(\mathrm{d}\mu_h)$. For $m \in M$ and $\varphi \in C(X)$ we define

$$U_\varphi m = V^{-1} \varphi V m.$$

Since $\chi_E L^2(\mathrm{d}\mu_h)$ is an invariant subspace to the multiplication by functions of $C(X)$, U_φ are well defined and linear on M. Obviously

$$\|U_\varphi\| \leqslant \|\varphi\|.$$

At the same time

$$U_{\varphi_1} U_{\varphi_2} m = V^{-1} \varphi_1 V V^{-1} \varphi_2 V m = V^{-1} \varphi_1 \varphi_2 V_m = U \varphi_1 \varphi_2 m.$$

Hence $\varphi \to U_\varphi$ is a representation of $C(X)$ on M.

From (8.3.16) there results

$$T_f m = V^{-1} f V m = U_f m \qquad (f \in A;\ m \in M).$$

For $m \in M$ and $k \in M_h$ we also have

$$(V_f^* m, k) = (m, V_f k) = (m, T_f k) = (m, V^{-1} f V k) =$$

$$= (Vm, fVk) = (\bar{f} Vm, Vk) = (V^{-1} \bar{f} Vm, k) = (U_f^* m, k).$$

We used (8.3.16) and the fact that V and V^{-1} are isometries.

Therefore M is doubly invariant to $f \to V_f$ and $\varphi \to U_\varphi$ is an extension to $C(X)$ of $f \to V_f$, i.e. the subspace M is X-spectral relative to $f \to V_f$.

Conversely, assume M is X-spectral relative to $f \to V_f$ and let $\varphi \to U_\varphi$ be the representation of $C(X)$ on M, which extends $f \to V_f|M$. According to (8.3.16), VM is invariant to multiplication by functions of A.

Let P be the projection of $L^2(d\mu_h)$ on VM.

For $m, n \in M$ we have

$$(V_f^* m, n) = (m, V_f n) = (m, T_f n) = (m, V^{-1} f V n) =$$

$$= (Vm, fVn) = (\bar{f} Vm, Vn) = (P\bar{f}Vm, Vn) =$$

$$= (V^{-1} P \bar{f} Vm, n).$$

Hence

$$V_f^* m = V^{-1} P \bar{f} V m. \tag{8.3.17}$$

Since V_f are normal operators, we have

$$\|P\bar{f}Vm\| = \|V^{-1}P\bar{f}Vm\| = \|V_f^* m\| = \|V_f m\| =$$

$$= \|V^{-1} f V m\| = \|fVm\| = \|\bar{f}Vm\|$$

for any $m \in M$. Therefore $\bar{f}VM \subset VM$ for any $f \in A$. Then VM is invariant to multiplication by any function of $C(X)$ of the form $\varphi = \Sigma f_i \bar{g}_i$ and, by continuity, to multiplication by any function $\varphi \in C(X)$. From Theorem 7.14 there results $VM = \chi_E L^2(d\mu_h)$, with E a Borel set, μ_h-essentially unique.

The proof is therefore complete. ◇

Theorem 8.19. *Let $f \to T_f$ be a subspectral representation of A on H and μ_h the spectral measure attached to h by $f \to T_f$. μ_h is a Szegö measure if and only if the representation $f \to T_f|M_h$ of A on M_h is completely nonspectral.*

Proof. Since, obviously, $V(M_h) \subset H^2(d\mu_h)$ then, according to Proposition 8.18, $f \to T_f|M_h$ has no X-spectral subspaces different from zero, if and only if for any Borel set E with $\chi_E L^2(d\mu_h) \subset H^2(d\mu_h)$ we have $\mu_h(E) = 0$, that is, if and only if μ_h is a Szegö measure.

Let μ be a positive measure on X with $\mu(X) = 1$ and $f \to T_f$ the natural representation of A on $H^2(\mathrm{d}\mu)$. Taking as $h \in H^2(\mathrm{d}\mu)$ the function identically equal to 1 on X, we have

$$\mu_1(\varphi) = (U_\varphi 1, 1) = (\varphi, 1) = \int \varphi \, \mathrm{d}\mu$$

hence μ is the spectral measure attached, by the natural representation, to the element 1 of $H^2(\mathrm{d}\mu)$.

Furthermore $M_1 = H^2(\mathrm{d}\mu)$ and the operator V defined above is the embedding operator of $H^2(\mathrm{d}\mu)$ in $L^2(\mathrm{d}\mu)$.

Theorem 8.20. *Let μ be a positive measure on X and $f \to T_f$ the natural representation of A on $H^2(\mathrm{d}\mu)$.*

(i) *The subspace M of $H^2(\mathrm{d}\mu)$ is X-spectral if and only if $M = \chi_E L^2(\mathrm{d}\mu)$, where E is a Borel set of X. E is μ-essentially unique determined by M.*

(ii) *The natural representation of A on $H^2(\mathrm{d}\mu)$ is completely non-spectral if and only if μ is a Szegö measure.*

(iii) *The representation $f \to T_f$ is spectral if and only if μ is Szegö-singular.*

(iv) *The measure μ has a unique decomposition under the form $\mu = \mu_1 + \mu_2$ where μ_1 is a Szegö-singular measure and μ_2 a Szegö measure. Moreover*

$$H^2(\mathrm{d}\mu) = H^2(\mathrm{d}\mu_1) \oplus H^2(\mathrm{d}\mu_2)$$

where the subspaces $H^2(\mathrm{d}\mu_1)$, $H^2(\mathrm{d}\mu_2)$ are doubly invariant, $f \to T_f | H^2(\mathrm{d}\mu_1)$ is the spectral part of $f \to T_f$, $f \to T_f | H^2(\mathrm{d}\mu_2)$ is the completely non-spectral part of $f \to T_f$. The measure μ_2 is the supremum of the Szegö measures ν, such that $\nu \leqslant \mu$.

Proof. (i), (ii) and (iii) are direct consequence of Proposition 8.18 and Theorem 8.19.

It remains to prove (iv). Let

$$H^2(\mathrm{d}\mu) = H_s \oplus H_c$$

the canonical decomposition of the representation $f \to T_f$. Since H_s is an X-spectral subspace, then, according to (i) it has the form

$$H_s = \chi_E L^2(\mathrm{d}\mu)$$

with E a Borel set in X. Let $\mu_1 = \chi_E \mu$. We have

$$L^2(d\mu_1) = \chi_E L^2(d\mu) = H_s \subset H^2(d\mu).$$

Since any function of $L^2(d\mu)$ which belongs to $H^2(d\mu)$, belongs also to $H^2(d\mu_1)$, there results $L^2(d\mu_1) = H_s = H^2(d\mu_1)$, hence μ is a Szegö-singular measure.

Let $\mu_2 = (1 - \chi_E)\,\mu$. Since $\chi_E \in H^2(d\mu)$, there follows

$$H_c = (1 - \chi_E)\,H^2(d\mu) = H^2(d\mu_2).$$

Then

$$H^2(d\mu) = H^2(d\mu_1) \oplus H^2(d\mu_2).$$

The representation $f \to T_f|H_c$ is therefore the natural representation of A on $H^2(d\mu_2)$; as it is completely non-spectral, μ_2 results a Szegö measure. Hence

$$\mu = \mu_1 + \mu_2 \tag{8.3.18}$$

with μ_1 a Szegö-singular measure and μ_2 a Szegö measure.

Let ν be a Szegö measure, $\nu \leqslant \mu$. We have

$$\chi_E C(X) \subset H^2(d\mu) \subseteq H^2(d\nu)$$

and therefore $\nu(E) = 0$, i.e. $\nu \leqslant \mu_2$. Thus, μ_2 is the supremum of the Szegö measures ν, $\nu \leqslant \mu$. This proves also the uniqueness of the decomposition (8.3.18).

The Theorem is proved. ◇

The measure μ_1 is called *the Szegö singular part* of μ, and μ_2 *the Szegö part of* μ.

8.4. The Wold decomposition

Let A be a function algebra on X and μ a representing measure for A. Recall we denoted

$$A_\mu = \{f \in A : \textstyle\int f d\mu = 0\}.$$

In the following we suppose that μ has the uniqueness property, i.e. any representing measure which coincides with μ on A is equal to μ.

The subspectral representation $f \to T_f$ of A on the Hilbert space H is said to be *μ-spectral* if

$$\operatorname{clm}\,[\bigcup_{f \in A_\mu} T_f H] = H.$$

Proposition 8.21. *Let $f \to T_f$ be a μ-spectral representation of A on H and $M \subset H$ a doubly invariant subspace. The representation $f \to T_f|M$ is a μ-spectral representation of A on M.*

Proof. According to Proposition 8.16, $f \to T_f|M$ is subspectral. Let $m \in M$, orthogonal to $[\bigcup_{f \in A_\mu} T_f M]$. For any $f \in A_\mu$ and $h \in H$ of the form $h = h_1 + h_2$, with $h_1 \in M$, $h_2 \in M^\perp$, we have

$$(m, T_f h) = (m, T_f h_1) + (m, T_f h_2) = 0.$$

Hence m is orthogonal to

$$\operatorname{clm}\,[\bigcup_{f \in A_\mu} T_f H]$$

which is equal to H, therefore $m = 0$. ◇

The subspectral representation $f \to T_f$ is called *μ-completely non spectral* if for any doubly invariant subspace M, for which $f \to T_f|M$ is μ-spectral, we have $M = \{0\}$.

Theorem 8.22. *Let $f \to T_f$ be a subspectral representation of A on H and μ a representing measure of A with the uniqueness property. The space H has a unique decomposition of the form*

$$H = H_1 \oplus H_2$$

such that H_1, H_2 are doubly invariant subspaces for $f \to T_f$, $f \to T_f|H_1$ is a μ-spectral representation and $f \to T_f|H_2$ is a μ-completely nonspectral representation. H_1 is the largest doubly invariant subspace for which $f \to T_f|M$ is a μ-spectral representation.

Proof. Let us write

$$N = [\bigcup_{f \in A_\mu} T_f A]^\perp.$$

Let $\varphi \to U_\varphi$ be the spectral dilation of $f \to T_f$ and $(\mu_{k_1, k_2})_{k_1, k_2 \in K}$ the spectral family attached to $\varphi \to U_\varphi$ (K is the space of spectral dilation). For any $n \in N$ and $f \in A_\mu$ we have:

$$\mu_n(f) = (U_f n, n) = (T_f n, n) = 0$$

according to the definition of N. Since

$$\mu_n(1) = \|n\|^2$$

using the uniqueness property of μ, there results

$$\mu_n = \|n\|^2 \mu \qquad (n \in N). \tag{8.4.1}$$

Let us put

$$H_1 = \text{clm}\,[\bigcup_{f \in A} T_f N].$$

It is clear that $T_f H_1 \subset H_1$ for any $f \in A$. If $n \in N$ and $g \in A_\mu$, then for any $h \in H$ we have:

$$(T_g^* n, h) = (n, T_g h) = 0$$

hence

$$T_g^* n = 0 \qquad (n \in N,\ g \in A_\mu). \tag{8.4.2}$$

It results that $T_g^* N \subset N$ for any $g \in A$. Let now $h \in H_1$ be of the form $h = T_f n$ with $n \in N$ and $f \in A$. For any $g \in A$ we can choose a sequence $f_k + \bar{g}_k$ with $f_k, g_k \in A$ such that

$$\lim_{k \to \infty} \int |f_k + \bar{g}_k - f\bar{g}|^2 \mathrm{d}\mu = 0,$$

since $A+\bar{A}$ is dense in $L^2(\mathrm{d}\mu)$ (Corollary 5.19). Since

$$\|(T_{f_k}+T^*_{g_k}-T^*_gT_f)n\|^2 \leqslant \|U_{f_k+\bar{g}_k-f\bar{g}}n\|^2 =$$

$$=\int|f_k+\bar{g}_k-f\bar{g}|^2\mathrm{d}\mu_n = \|n\|^2\int|f_k+\bar{g}_k-f\bar{g}|^2\mathrm{d}\mu$$

we obtain

$$\lim_{k\to\infty}(T_{f_k}n+T^*_{g_k}n)=T^*_gT_fn.$$

Hence

$$T^*_gh=T^*_gT_fn=\lim_{k\to\infty}(T_{f_k}n+T^*_{g_k}n)\in H_1$$

thus the subspace H_1 is doubly invariant. Let us write

$$H=H_1\oplus H_2$$

where H_1 and H_2 are doubly invariant subspace. We have

$$\operatorname{clm}[\bigcup_{f\in A_\mu}T_fH_1]=H_1.$$

Indeed if $m\in H_1$ is orthogonal to

$$\operatorname{clm}[\bigcup_{f\in A_\mu}T_fH_1]$$

then for any $f\in A_\mu$ and $h\in H$ of the form $h=h_1+h_2$ with $h_1\in H_1$, $h_2\in H_2$ we have

$$(m,T_fh)=(m,T_fh_1)+(m,T_fh_2)=0$$

therefore m is orthogonal to

$$\operatorname{clm}[\bigcup_{f\in A_\mu}T_fH]$$

which contains H_1, i.e. $m=0$.

It results that $f \to T_f|H_1$ is μ-spectral.

Let M be a doubly invariant subspace of H such that

$$\text{clm}\,[\bigcup_{f\in A_\mu} T_f M] = M.$$

Then it is easy to see that M is orthogonal to N. Then for any $f \in A$, $m \in M$ and $n \in N$ we have

$$(m, T_f n) = (T_f^* m, n) = 0.$$

Hence M is orthogonal to H_2 i.e. $M \subset H_1$. Therefore H_1 is the maximal doubly invariant subspace M for which $f \to T_f|M$ is μ-spectral. It is now clear that $f \to T_f|H_2$ is μ-completely non-spectral.

Let

$$H = H_1' \oplus H_2'$$

be another decomposition of H as required by the theorem. By the maximality of H_2 we have $H_1' \subset H_1$. Let us put

$$H_1 = H_1' \oplus M.$$

It is then clear that M is a doubly invariant subspace and, since $M \subset H_1$, from Proposition 8.21 there results that $f \to T_f|M$ is μ-spectral. Since $M \subset H_2'$ we conclude that $M=\{0\}$. Thus $H_1' = H_1$, $H_2' = H_2$. $\diamond$

Proposition 8.23. *If μ is not a point measure and M is a spectral subspace of $f \to T_f$ then $f \to T_f|M$ is μ-spectral.*

Proof. Let

$$N = [\bigcup_{f\in A_\mu} T_f H]^\perp$$

and assume there is an $h \in M \cap N$, $\|h\| = 1$. Let M_h be the closed subspace generated by the elements of the form $T_f h$, $f \in A$. As in the case of H_2 from the previous theorem, we can prove that M_h is doubly invariant to $f \to T_f$. As $M_h \subset M$ and M is spectral, M_h is also spectral and, according to Proposition 8.18, there exists a Borel set E, $\mu_h =$ $= \mu$ — essentially unique, such that

$$VM_h = \chi_E L^2(\mathrm{d}\mu)$$

where V is the operator defined on M_h, with values in $L^2(\mathrm{d}\mu)$, by

$$V(T_f h) = f.$$

But, clearly, $VM_h \subset H^2(\mathrm{d}\mu)$, and therefore $\chi_E L^2(\mathrm{d}\mu) \subset H^2(\mathrm{d}\mu)$. According to Theorem 8.14, μ is a Szegö measure, hence $\mu(E) = 0$. There results $M_h = 0$ which contradicts the fact that $h \in M_h$ and $\|h\| = 1$. Therefore

$$M \cap N = \{0\}.$$

Now, let $m \in M$ be orthogonal to

$$\mathrm{clm}\,[\bigcup_{f \in A_\mu} T_f M].$$

For any $h \in H$ of the form $h = h_1 + h_2$, with $h_1 \in M$, $h_2 \in M^\perp$ and $f \in A_\mu$ we have

$$(m, T_f h) = (m, T_f h_1) + (m\ T_f h_2) = 0.$$

This means that $m \in [\bigcup_{f \in A_\mu} T_f H]^\perp = N$, hence $m \in M \cap N = \{0\}$, i.e. $m = 0$. ◇

Corollary 8.24. *Let $f \to T_f$ be a subspectral representation of A on H and μ a measure which is not a point measure. If $f \to T_f$ is spectral, then it is also μ-spectral. If $f \to T_f$ is μ-completely non-spectral, then it is also completely nonspectral.*

A subspectral representation $f \to T_f$ of A on H, μ-spectral and completely non-spectral, will be called *μ-singular*.

Theorem 8.25. (Wold-Helson-Lawdenslager).. *Let $f \to T_f$ be a subspectral representation of A on H and μ a representing measure (which* *is not a point measure) for A with the uniqueness property. H admits a unique decomposition under the form*

$$H = H_1 \oplus H_2 \oplus H_3$$

such that H_1, H_2, H_3 *are doubly invariant subspaces to* $f \to T_f, f \to T_f|H_1$ *is spectral,* $f \to T_f|H_2$ *is* μ*-singular and* $f \to T_f|H_3$ *is* μ*-completely non-spectral.*

Proof. The proof follows immediately by applying successively Theorem 8.6, Corollary 8.24 and Theorem 8.22. ◇

The following theorem furnishes some characterizations of natural representations and gives the reason for which Theorem 8.24 was called the Wold Theorem.

Theorem 8.26. *Let* $f \to T_f$ *be a subspectral representation of A on H and* μ *a non-point mass representing measure of A with the uniqueness property. The following assertions are equivalent.*

a) *There exists a Hilbert space* N *such that* $f \to T_f$ *is unitarily equivalent to the natural representation of A on* $H^2(\mathrm{N}; \mathrm{d}\mu)$.

b) *The representation* $f \to T_f$ *is* μ*-completely non-spectral*

c) *If M is a closed subspace of H such that* $T_f^* M \subset M$ *for any* $f \in A$ *and*

$$M \subset \operatorname{clm}\,[\bigcup_{f \in A_\mu} T_f H],$$

then $M = \{0\}$.

d) *If M is a doubly invariant subspace of H with*

$$M \subset \operatorname{clm}\,[\bigcup_{f \in A_\mu} T_f H]$$

then $M = \{0\}$.

e) *If M is a doubly invariant subspace such that*

$$M \cap [\bigcup_{f \in A_\mu} T_f H]^\perp = \{0\}$$

then $M = \{0\}$.

Proof. Let us prove the implications a) → c). For this, let $f \to T_f$ be the natural representation of A on $H^2(N; d\mu)$. First we shall show that

$$N = [\bigcup_{f \in A_\mu} T_f H^2(N; d\mu)]^\perp. \tag{8.4.3}$$

Indeed if $n \in N$, $f \in A_\mu$, $g \in A$ and $m \in N$, we have:

$$(T_f gm, n) = (fgm, n) = \int (f(x)\, g(x)\, m, n)\, d\mu(x) =$$

$$= (m, n) \int fg\, d\mu = 0.$$

Conversely, if $h \in H^2(N; d\mu)$ is orthogonal to N and

$$[\bigcup_{f \in A_\mu} T_f H^2(N; d\mu)],$$

then for any $n \in N$ and $g \in A$, if we put

$$f = g - \int g d\mu,$$

we obtain

$$(h, gn) = (h, fn) + (h, \int g d\mu\, n) =$$

$$= (h, T_f n) + \int g d\mu\, (h, n) = 0.$$

Let now M be a closed subspace of $H^2(N; d\mu)$ as in *c*). From (8.4.3) M results orthogonal to N. Then for any $m \in M$, $f \in A$ and $n \in N$ we have

$$(m, fn) = (m, T_f n) = (T_f^* m, n) = 0$$

i.e. $m = 0$. Hence $M = \{0\}$ and implication a) → c) is proved.

c) → d) and d) → b) are obvious.

b) → e). Let M be a doubly invariant subspace as in e). Then we have

$$\text{clm}\, [\bigcup_{f \in A_\mu} T_f M] = M. \tag{8.4.4}$$

Indeed, if $m \in M$ is orthogonal to

$$[\bigcup_{f \in A_\mu} T_f M]$$

then for any $h \in H$ of the form $h = h_1 + h_2$, with $h_1 \in M$, $h_2 \in M^\perp$, and $f \in A$ we have

$$(m, T_f h) = (m, T_f h_1) + (m, T_f h_2) = 0.$$

Thus

$$m \in [\bigcup_{f \in A_\mu} T_f H]^\perp \cap M$$

i.e. $m = 0$. Hence M verifies (8.4.4) and from b) we obtain $M = \{0\}$. This completes the proof b) → e).

e) → d) is obvious.

To complete the proof of the theorem it remains to prove b) → a). If b) is true then from Theorem 8.22 there results that

$$H = \text{clm}\,[\bigcup_{f \in A} T_f N]$$

where

$$N = [\bigcup_{f \in A_\mu} T_f H]^\perp.$$

Let $(\mu_{k_1, k_2})_{k_1, k_2 \in K}$ be the spectral family attached to the spectral dilation $\varphi \to U_\varphi$ of $f \to T_f$. Using (7.1.3) and (8.4.1), for any $n, m \in N$, We obtain

$$4\mu_{n,m} = \mu_{n+m} - \mu_{n-m} + i\mu_{n+im} - i\mu_{n-im} =$$

$$=(\|n+m\|^2 - \|n-m\|^2 + i\|n+im\|^2 - i\|n-im\|^2) = 4(n, m)\mu.$$

Then for any $h \in H$ of the form $\sum_{i=1} T_{f_i} n_i$, with $f_i \in A$ and $n_i \in N$, we have:

$$\left\| \sum_{i=1}^{p} T_{f_i} n_i \right\|^2 = \left(\sum_{i=1}^{p} T_{f_i} n_i, \sum_{j=1}^{p} T_{f_j} n_j \right) =$$

$$= \left(\sum_{i=1}^{p} U_{f_i} n_i, \sum_{i=1}^{p} U_{f_j} n_j \right) = \sum_{i,j=1}^{p} (U_{f_i \bar{f}_j} n_i, n_j) =$$

$$= \sum_{i,j=1}^{p} \int f_i \bar{f}_j \mathrm{d}\mu_{n_i n_j} = \sum_{i,j=1}^{p} \int f_i \bar{f}_j (n_i, n_j) \,\mathrm{d}\mu = \int \sum_{i,j=1}^{p} f_i \bar{f}_j (n_i, n_j) \,\mathrm{d}\mu =$$

$$= \int \left\| \sum_{i=1}^{p} f_i n_i \right\|^2 \mathrm{d}\mu = \left\| \sum_{i=1}^{p} f_i n_i \right\|^2 .$$

Now it is clear that the map $T_f n \to fn$ may be extended to an isometric operator U from H on $H^2(N; \mathrm{d}\mu)$ such that

$$UT_f h = fUh \qquad (f \in A,\ h \in H).$$

This completes the proof of the theorem. ◇

The following theorem completes Theorem 8.26 in the case of one-dimensional N.

Theorem 8.27. *The subspectral representation $f \to T_f$ of A on H is unitarily equivalent to the natural representation of A on H^2 $(N; \mathrm{d}\mu)$ with N-one-dimensional, if and only if it is not μ-spectral and has no proper doubly invariant subspaces.*

Proof. Let $f \to T_f$ be the natural representation of A on $H^2(N; \mathrm{d}\mu)$ with N one-dimensional, and M a doubly invariant subspace. If $M \cap N = 0$ then from Theorem 8.2.6 there results $M = \{0\}$. If $M \cap N \neq 0$ then $M \cap N = N$ since N is one-dimensional, hence $M = H^2(N; \mathrm{d}\mu)$.

Now, let $f \to T_f$ be a subspectral representation which is not μ-spectral and has no doubly invariant subspaces. Since

$$H \bigcap [\bigcup_{f \in A_\mu} T_f H]^\perp \neq 0,$$

according to Theorem 8.26 point e), there exists a Hilbert space N such that $f \to T_f$ is unitarily equivalent to the natural representation of $H^2(N; d\mu)$. Let $N_0 \subset N$ be one-dimensional and M be the closed subspace spanned by the elements of the form fn, $f \in A$, $n \in N$. Repeating the arguments used in Theorem 8.22 for proving that H_2 is doubly invariant, we can show that M is a doubly invariant subspace. Since $M \neq 0$ we have $H = M = H^2(N; d\mu)$. ◇

8.5. Decomposition with respect to Gleason parts

Let A be a Dirichlet algebra on X and $\{G_\alpha\}_{\alpha \in J}$ the family of Gleason parts of the maximal ideal space of A. For any $\alpha \in J$ let m_α be a representing measure for an element φ_α of G_α. Since A is a Dirichlet algebra on X, m_α is the unique representing measure, with support in X, for φ_α. According to Proposition 6.5, for $\alpha \neq \beta$, m_α, m_β are mutually singular.

Let H be a Hilbert space and $f \to T_f$ a representation of A on H. As A is a Dirichlet algebra, there exists a unique spectral dilation $\varphi \to U_\varphi$ of $f \to T_f$, hence there is a unique semispectral family $(\mu_{h,k})_{h,k \in H}$ on X, such that

$$(T_f h, k) = \int f d\mu_{h,k} \qquad (f \in A; h, k \in H).$$

If P is the orthogonal projection of the dilation space on H then, $\mu_{h,k}$ are defined by

$$\mu_{h,k}(g) = (PU_g h, k) \qquad (g \in C(X);\ h, k \in H).$$

We write, as usual, $\mu_h = \mu_{h,h}$.

The representation $f \to T_f$ is called *G_α-continuous* if for any $h \in H$, μ_h is absolutely continuous with respect to m_α. The representation $f \to T_f$ will be called *singular* if for any $\alpha \in J$ and $h \in H$, m_α and μ_h are mutually singular measures.

Following Theorem 3.12, if $\varphi_1, \varphi_2 \in G_\alpha$ and m_1, m_2 are their representing measures, with support in X, then m_1 and m_2 are mutually absolutely continuous. Therefore, the notions of G_α-continuity and singularity do not depend on the particular choice of the elements φ_α in G_α.

Theorem 8.28. *Let A be a Dirichlet algebra on X and $\{G_\alpha\}_{\alpha \in J}$ the family of Gleason parts of the maximal ideal space of A. Let H be a Hilbert space and $f \to T_f$ a representation of A on H. Then, H has a unique decomposition under the form*

$$H = \bigoplus_{\alpha \in J} H_\alpha \oplus H_0$$

where H_α, $\alpha \in J$, and H_0 are doubly invariant subspaces, $f \to T_J|H_\alpha$, $\alpha \in J$, is G_α-continuous and $f \to T_f|H_0$ is singular.

Proof. Let $(\mu_{h,k})_{h,k \in H}$ be the semispectral family attached to the representation $f \in T_f$. For any $\alpha \in J$ we fix the representing measure m_α of an element φ_α in G_α.

Let

$$\mathrm{d}\mu_{h,k} = m^a_{h,k} + \sigma^a_{h,k} \qquad (h, k \in H)$$

be the Lebesgue decomposition of the measure $\mu_{h,k}$ with respect to m_α; $m^a_{h,k}$ is the absolutely continuous part of $\mu_{h,k}$ relative to m_α and $\sigma^a_{h,k}$ the singular part.

Since $(\mu_{h,k})_{h,k \in H}$ is a semispectral family on H, one easily verifies that $(m^a_{h,k})_{h,k \in H}$ and $(\sigma^a_{h,k})_{h,k \in H}$ are also semispectral families on H, for any $\alpha \in J$.

Let $\varphi \to T^a_\varphi$ be the positive definite map of $C(X)$ in $L(H)$ attached to the semispectral family $(m^a_{h,k})_{h,k \in H}$

$$(T^a_\varphi h, k) = \int \varphi \, \mathrm{d}m^a_{h,k} \qquad (\varphi \in C(X);\ h, k \in H).$$

We have

$$(T_{fg}h, k) = \int fg \, \mathrm{d}\mu_{h,k} = \int fg \, \mathrm{d}m^a_{h,k} + \int fg \, \mathrm{d}\sigma^a_{h,k}$$

$$(T_{fg}h, k) = (T_f T_g h, k) = \int f \mathrm{d}\mu_{T_g h, k} = \int f \mathrm{d}m^a_{T_g h, k} +$$

$$+ \int f \mathrm{d}\sigma^a_{T_g h, k}$$

$$(T_{fg}h, k) = (T_f T_g h, k) = (T_g h, T_f^* k) = \int g \, \mathrm{d}\mu_{h, T_f^* k} =$$

$$= \int g \, \mathrm{d}m^a_{h, T_f^* k} + \int g \, \mathrm{d}\sigma^a_{h, T_f^* k}$$

for $f, g \in A$. Therefore, the measures

$$f\,\mathrm{d}m^{a}_{h,k} + f\,\mathrm{d}\sigma^{a}_{h,k} - \mathrm{d}\,m^{a}_{h,T^{*}_{f}k} - \mathrm{d}\sigma^{a}_{h,T^{*}_{f}k}$$

$$g\,\mathrm{d}\,m^{a}_{h,k} + g\,\mathrm{d}\sigma^{a}_{h,k} - \mathrm{d}\,m^{a}_{T_g h,k} - \mathrm{d}\sigma^{a}_{T_g h,k}$$

are orthogonal to A. According to *F.* and M. Riesz Theorem (Theorem 5.6) the absolutely continuous and singular parts of these measures, relative to m_α, are orthogonal to A too. Then, we have

$$\text{(8.5.1)} \qquad \begin{aligned} \int fg\,\mathrm{d}m^{a}_{h,k} &= \int g\,\mathrm{d}m^{a}_{h,T^{*}_{f}k} \\ \int fg\,\mathrm{d}\sigma^{a}_{h,k} &= \int g\,\mathrm{d}\sigma^{a}_{h,T^{*}_{f}k} \end{aligned} \qquad (f, g \in A;\ h, k \in H)$$

and

$$\text{(8.5.2)} \qquad \begin{aligned} \int fg\,\mathrm{d}m^{a}_{h,k} &= \int f\,\mathrm{d}m^{a}_{T_g h,k} \\ \int fg\,\mathrm{d}\sigma^{a}_{h,k} &= \int f\,\mathrm{d}\sigma^{a}_{T_g h,k} \end{aligned} \qquad (f, g \in A;\ h, k \in H).$$

From (8.5.1) we get

$$\int f\,\mathrm{d}m^{a}_{T^{a}_{g}h,k} + \int f\,\mathrm{d}\sigma^{a}_{T^{a}_{g}h,k} - \int fg\,\mathrm{d}m^{a}_{h,k} =$$

$$= \int f\,\mathrm{d}\mu_{T^{a}_{g}h,k} - \int fg\,\mathrm{d}m^{a}_{h,k} = (T_f T^{a}_{g}h, k) - \int fg\,\mathrm{d}m^{a}_{h,k} =$$

$$= (T^{a}_{g}h, T^{*}_{f}k) - \int fg\,\mathrm{d}m^{a}_{h,k} = \int g\,\mathrm{d}m^{a}_{h,T^{*}_{f}k} - \int fg\,\mathrm{d}m^{a}_{h,k} = 0$$

for any $f, g \in A$. Hence, the measure

$$\mathrm{d}m^{a}_{T^{a}_{g}h,k} + \mathrm{d}\sigma^{a}_{T^{a}_{g}h,k} - g\,\mathrm{d}m^{a}_{h,k}$$

is orthogonal to A and, F. and M. Riesz Theorem yields

$$\text{(8.5.3)} \qquad \int f\,\mathrm{d}m^{a}_{T^{a}_{g}h,k} = \int g\,\mathrm{d}m^{a}_{h,k} \qquad (f, g \in A;\ h, k \in H).$$

Now from (8.5.3) there results

$$(T_f^a T_g^a h, k) = \int f \,\mathrm{d}m^a_{T_g^a h, k} = \int fg \,\mathrm{d}m^a_{h,k} = (T^a_{fg} h, k)$$

for any $f, g \in A$, $h, k \in H$, that is

$$T_f^a T_g^a = T^a_{fg} \qquad (f, g \in A). \tag{8.5.4}$$

Therefore, the restriction of $\varphi \to T^a_\varphi$ to A is multiplicative. In particular $T_1^a = P_\alpha$ is a projection on H.

Using (8.5.1) and (8.5.3) we obtain

$$(T_f P_\alpha h, k) = (T_f T_1^a h, k) = (T_1^a h, T_f^* k) = \int \mathrm{d}m^a_{h, T_f^* k} =$$

$$= \int \mathrm{d}m^a_{h,k} = (T_f^a h, k)$$

$$(P_\alpha T_f h, k) = (T_1^a T_f h, k) = \int \mathrm{d}m^a_{T_f h, k} = \int f \,\mathrm{d}m^a_{h,k} = (T_f^a h, k)$$

which yields

$$P_\alpha T_f = T_f P_\alpha = T_f^a \qquad (f \in A) \tag{8.5.5}$$

$$P_\alpha T_f^a = T_f^a P_\alpha = T_f^a \qquad (f \in A). \tag{8.5.6}$$

If we put $H_\alpha = P_\alpha H$, then from (8.5.5) and (8.5.6) H_α results doubly invariant to $f \to T_f$ and $f \to T_f^a$, and $T_f | H_\alpha = T_f^a | H_\alpha$. It is also clear that $f \to T_f^a | H_\alpha$ is a G_α-continuous representation of A on H_α.

We now show that $P_\alpha P_\beta = 0$ for $\alpha \neq \beta$.

We have

$$\int f \,\mathrm{d}m^a_{T_1^\beta h, k} + \int f \,\mathrm{d}\sigma^a_{T_1^\beta h, k} - \int f \,\mathrm{d}m^\beta_{h,k} = \int f \,\mathrm{d}\mu_{T_1^\beta h, k} - \int f \,\mathrm{d}m^\beta_{h,k} =$$

$$= (T_f T_1^\beta h, k) - \int f \,\mathrm{d}m^\beta_{h,k} = (T_1^\beta h, T_f^* k) - \int f \,\mathrm{d}m^\beta_{h,k} =$$

$$= \int \mathrm{d}m^\beta_{h, T_f^* k} - \int f \,\mathrm{d}m^\beta_{h,k} = 0.$$

From (8.5.1), the measure

$$m^{\alpha}_{T_1^{\beta}h,\,k} + \sigma^{\alpha}_{T_1^{\beta}h,\,k} - m^{\beta}_{h,\,k}$$

results orthogonal to A. Then, using again F. and M. Riesz Theorem and taking into account that m_α and m_β are mutually singular, we get

$$\int f \, dm^{\alpha}_{T_1^{\beta}h,\,k} = 0 \qquad (f \in A). \tag{8.5.7}$$

Hence

$$(P_\alpha P_\beta\, h, k) = (T_1^{\alpha} T_1^{\beta} h,\; k) = \int dm^{\alpha}_{T_1^{\beta}h,\,k} = 0$$

for any $h, k \in H$, that is

$$P_\alpha P_\beta = 0.$$

Therefore the subspaces H_α are pairwise orthogonal. Let

$$H = \bigoplus_{\alpha \in \mathscr{I}} H_\alpha \oplus H_0 .$$

As each H_α is doubly invariant to $f \to T_f$, H_0 results doubly invariant to $f \to T_f$ too. We know that $f \to T_f | H_\alpha$, $\alpha \in \mathscr{I}$ is G_α-continuous. Let us prove that $f \to T_f | H_0$ is singular. First we observe that from (8.5.3) there results

$$\int dm^{\alpha}_{h,\,h} = \int dm^{\alpha}_{T_1^{\alpha}h,\,h} = \int dm^{\alpha}_{P_\alpha h, h} = \int dm^{\alpha}_{0,h} = 0$$

for any $h \in H_0$ and, as m_h is a positive measure, $m_h = 0$ for any $h \in H_0$. This means that $f \to T_f | H_0$ is singular.

There remained to prove that the decomposition is unique. Let

$$H = \bigoplus_{\alpha \in \mathscr{I}} H'_\alpha \oplus H'_0$$

be another decomposition of H as required in the theorem, and let P'_a, P'_0 be the orthogonal projection of H on H'_a, H'_0 respectively. Then

$$\mu_{P'_a h, P'_a k} = m^a_{P'_a h, P'_a k} \qquad (\alpha \in \mathscr{J}; h, k \in H)$$

(8.5.8)
$$m^a_{P'_0 h, P'_0 k} = 0.$$

At the same time, since H'_a are doubly invariant to $f \to T_f$, there follows

$$\int f d\mu_{h,k} = (T_f h, k) = 0$$
$$\int f d\mu_{k,h} = (T_f k, h) = 0$$

for any $h \in H'_a$, $k \in H'^{\perp}_a$ and $f \in A$.

Therefore $\mu_{h,k}$, $\mu_{k,h}$ are orthogonal to A, and applying F. and M. Riesz Theorem we obtain

(8.5.9) $\int f dm^a_{h,k} = \int f dm^a_{k,h} = 0 \quad (f \in A; h \in H'_a;\ k \in H'^{\perp}_a;\ \alpha \in \mathscr{J}).$

Using (8.5.8) we have

$$(P'_a P_\alpha P'_a h, k) = (P_\alpha P'_a h, P'_a k) = (T^a_1 P'_a h, P'_a k) = \int dm^a_{P'_a h, P'_a k} =$$
$$= \int d\mu_{P'_a h, P'_a k} = (P'_a h, P'_a k) = (P'_a h, k)$$

that is

(8.5.10)
$$P'_a P_\alpha P'_a = P'_a.$$

Now, from (8.5.9) and (8.5.10) we obtain

$$(P_\alpha P'_a h, k) = (P_\alpha P'_a h, P'_a k) + (P_\alpha P'_a h, (I - P'_a) k) =$$
$$= (P'_a P_\alpha P'_a h, k) + \int dm^a_{P'_a h, (I - P'_a) k} = (P'_a h, k)$$

$$(P'_a P_\alpha h, k) = (P_\alpha h, P'_a k) = (P_\alpha P'_a h, P'_a k) + (P_\alpha (I - P'_a) h, k) =$$
$$= (P'_a P_\alpha P'_a h, k) + \int dm^a_{(I - P'_a) h, P'_a k} = (P'_a h, k)$$

and therefore

$$(8.5.11) \qquad P_\alpha P'_\alpha = P'_\alpha = P'_\alpha P_\alpha.$$

From (8.5.11) we have

$$P'_\alpha P_\beta = P_\beta P'_\alpha = 0.$$

Quite similarly, one proves that

$$P_\alpha P'_0 = 0$$

which gives

$$P_\alpha = P_\alpha (P'_\alpha + \sum_{\beta \neq \alpha} P'_\beta + P'_0) = P_\alpha P'_\alpha = P'_\alpha.$$

The theorem is proved. ◇

Notes

Theorem 8.6 has been proved by C. Foiaş, I. Suciu [1]. The same work presents for the first time the notion of spectral dilation of a function algebra representation and the existence and uniqueness theorem of a spectral dilation for representation of Dirichlet algebras (Theorem 8.7).

Actually Theorem 8.13 is given in C. Foiaş and I. Suciu [2] in a form including Proposition 8.14. The present discussion in § 8.2 is self-contained, avoiding the theory of p-summing operators, which was used in the above quoted Note.

The Szegö measures have been introduced by C. Foiaş and I. Suciu [1]; the same paper contains the proofs to Theorems 8.15 and 8.16.

The connection between completely non-spectral representations and Szegö measures is thoroughly studied in W. Mlak [5].

Subspectral representations have been introduced by I. Suciu [10]. Theorem 8.20 has been proved for natural representations by C. Foiaş and I. Suciu [1]. In I. Suciu [10] it appears under the same form as presented here. μ-spectral, μ-singular and μ-completely non-spectral representations as well as the Wold type decomposition theorem for subspectral representations (Theorem 8.25) can be found in I. SUCIU [11]. The characterizations of natural representations contained in Theorem 8.26 have been proved in the same work.

G_α-continuous and singular representations have been introduced by W. MLAK [6]. In [6], W. MLAK proves Theorem 8.28 for any function algebra. Important contributions in this direction can also be found in D. GAŞPAR [2].

As concerns the problem of the existence of a spectral dilation for function algebras representations, a quite important result has been obtained in W. ARVESON [1], [2]; it is proved (as a particular case of a more general result for C*-algebras) that a representation $f \to T_f$ of A on H *has a spectral dilation if and only if it is completely contractive*, i.e. if and only if for any $n = 1, 2, \ldots$ and any $n \times n$ matrix (f_{ij}) over A, with $\|(f_{ij})\| \leqslant 1$, we have $\|(T_{f_{ij}})\| \leqslant 1$. The example given in chapter 10 (Corollary 10.11), based on PARROTT's example [1], shows that there exist representations of function algebras which are not completely contractive, hence have no spectral dilation.

CHAPTER 9

Elements of prediction theory on S-generated algebras

9.1. Semigroups of contractions

Let G be an abelian group. We write 1 for the unit element of G.

A *unitary representation* of G on the Hilbert space H, is a map $g \to U_g$ of G in $L(H)$ such that $U_1 = I$, $U_{g^{-1}} = U_g^*$ and $U_{g_1 g_2} = U_{g_1} U_{g_2}$ for any $g, g_1, g_2 \in G$. Obviously, U_g is a unitary operator on H for any $g \in G$.

Let $g \to T_g$ be a map of G in $L(H)$. *A unitary dilation* of the map $g \to T_g$ is a unitary representation of the group G on a Hilbert space K such that $H \subset K$ and

$$T_g h = PU_g h \qquad (h \in H, g \in G) \tag{9.1.1}$$

where P is the orthogonal projection of K on H. The unitary dilation is called *minimal* if the space K is the closure of the set spanned by the elements $U_g h$, $h \in H$, $g \in G$. In the following we shall assume every unitary dilation to be minimal.

If $g \to T_g$ admits a unitary dilation, then it is clear that $T_1 = I$, $T_{g^{-1}} = T_g^*$, $\|T_g\| \leqslant 1$ for any $g \in G$.

Theorem 9.1. (M. A. Naimark-B. Sz.-Nagy). *Let $g \to T_g$ be a map of G in $L(H)$, which satisfies $T_1 = I$, $T_{g^{-1}} = T_g^*$, for any $g \in G$. The map $g \to T_g$ admits a unitary dilation if and only if for any finite* *system $g_1,\ldots, g_n$ of elements in G and any finite system $h_1,\ldots, h_n$ of elements in H, we have*

$$\sum_{i,j} (T_{g_i^{-1} g_j} h_j, h_i) \geqslant 0. \tag{9.1.2}$$

Proof. Let $g \to U_g$ be a unitary dilation of $g \to T_g$. For any $g_1,\ldots, g_n \in G$ and $h_1, \ldots, h_n \in H$, we have

$$\sum_{i,j} (T_{g_i^{-1} g_j} h_j, h_i) = \sum_{i,j} (PU_{g_i^{-1} g_j} h_j, h_i) =$$

$$= \sum_{i,j} (U_{g_i^{-1} g_j} h_j, h_i) = \sum_{i,j} (U_{g_j} h_j, U_{g_i} h_i) =$$

$$= \| \sum_i U_{g_i} h_i \|^2 \geqslant 0.$$

Therefore (9.1.2) is a necessary condition for the existence of the unitary dilation of $g \to T_g$.

Now assume that $g \to T_g$ satisfies (9.1.2). For $h \in H$, we define on G the real function ρ_h, by

$$\rho_h(g) = (T_g h, h) \qquad (g \in G).$$

From (9.1.2) there results

$$\sum_{i,j} \rho_h(g_i^{-1} g_j) \bar{c}_i c_j = \sum_{i,j} (T_{g_i^{-1} g_j} c_j h, c_i h) \geqslant 0$$

for any finite system $c_1,\ldots, c_n$ of complex numbers. This means that ρ_h is a positive definite function on G and then, according to Herglotz-Bochner-Weil theorem, there exists a positive measure μ_h on the dual X of the discrete group G, such that

$$\rho_h(g) = \int g \, \mathrm{d}\mu_h \qquad (g \in G)$$

where g is considered as a character on X.

One easily verifies that $\|\mu_h\| = \|h\|$ and

$$\mu_{h_1+h_2} + \mu_{h_1-h_2} = 2\mu_{h_1} + 2\mu_{h_2} \qquad (h_1, h_2 \in H).$$

Writing

$$4\mu_{h,k} = \mu_{h+k} - \mu_{h-k} + i\mu_{h+ik} - i\mu_{h-ik} \qquad (h, k \in H),$$

by usual arguments one shows that $(\mu_{h,k})_{h,k\in H}$ is a semispectral family on X. Let $\varphi \to T_\varphi$ be the positive definite map attached to this family (Theorem 7.1) and $\varphi \to U_\varphi$ the spectral dilation of $\varphi \to T_\varphi$ (Theorem 7.5). It is easily verified that $g \to U_g$, is a unitary dilation of $g \to T_g$.

The proof is complete. ◇

The map $g \to T_g$ of G in $L(H)$, which verifies (9.1.2) will be called a *positive definite map*. From the proof of the preceding theorem there also results the following:

Proposition 9.2. *Let G be an abelian group and X the dual of the discrete group G. There exists a one-to-one correspondence between the set of positive definite maps $g \to T_g$ of G in $L(H)$, and the family of semispectral measures $(F(\sigma))_{\sigma\in B(X)}$ on X, given by the relation*

$$(T_g h, k) = \int g(x)\, d(F(x)h, k) \qquad (h, k \in H).$$

The map $g \to T_g$ is a unitary representation if and only if $(F(\sigma))_{\sigma\in B(X)}$ is a spectral measure.

Corollary 9.3. *Let G be an abelian group, X the dual of the discrete group G and $g \to T_g$ a positive definite map of G in $L(H)$. Then for any finite system of constants $c_1,\ldots, c_n$ and any finite system $g_1,\ldots, g_n$ of elements in G, the following inequality (Von Neumann formula)*

$$\| \Sigma c_i T_{ig} \| \leqslant \sup_{x\in X} | \Sigma c_i g_i(x) |$$

holds.

The uniqueness theorem for minimal dilations can be given as

Theorem 9.4. *Let $g \to T_g$ be a positive definite map of G in $L(H)$ and $g \to U_g$, $g \to U_g'$ two minimal unitary dilations acting on K and K',* *respectuvely. Let V, V' be the embedding operators of H in K and K', respectively. There exists a unitary operator $U: K \to K'$ such that*

$$UV = V'$$

$$UU_g = U'_g U \qquad (g \in G).$$

Proof. The proof follows from Proposition 9.2 and Theorem 7.6 ◇

We establish now a theorem which gives a sufficient condition for (9.1.2) to hold.

Theorem 9.5. *Let G be an abelian group, $S \subset G$ a subsemigroup of G with $S \cap S^{-1} = \{1\}$. Let $g \to T_g$ be a map of G in $L(H)$ such that $T_1 = I$, $T_{g^{-1}} = T_g^*$, $\|T_g\| \leqslant 1$ and*

$$T_{g_1} T_{g_2} = T_{g_1 g_2} \text{ for } g_1, g_2, g_1 g_2^{-1} \notin S^{-1}. \tag{9.1.3}$$

Then $g \to T_g$ satisfies (9.12).

Proof. Let $g_1, \ldots, g_n$ be a finite system of elements in G, with $g_i \neq g_j$ for $i \neq j$. If we choose minimal elements relative to the order induced by S in G, we may rearrange this system, such that

$$g_i^{-1} g_j \notin S^{-1} \text{ for any } \quad i < j. \tag{9.1.4}$$

Let $\widetilde{T} = (T_{ij})$ be the operator matrix given by $T_{ij} = T_{g_i^{-1} g_j}$. $\widetilde{T}$ is an operator acting on $\bigoplus_{k=1}^{n} H^k$(with $H^k = H$ for any k).

If $\widetilde{h} = (h_k)$ is a vector in $\bigoplus_{k=1}^{n} H^k$, then

$$(\widetilde{T}\,\widetilde{h}, \widetilde{h}) = \sum_{i,j=1}^{n} (T_{g_i^{-1} g_j} h_j, h_i)$$

hence (9.1.2) is equivalent to the positivity of the operator $\widetilde{T}$.

We show that $\widetilde{T} = \widetilde{W}^* \widetilde{D} \widetilde{W}$, where $\widetilde{W}$ and $\widetilde{D}$ are operators on $\bigoplus_{k=1}^{n} H^k$ and $\widetilde{D}$ is positive.

We put

$$W_{ij} = \begin{cases} T_{g_i^{-1} g_j} & \text{if } i \leqslant j \\ 0 & \text{if } i > j \end{cases}$$

and

$$D_{11} = I$$

$$D_{ij} = 0 \quad \text{if} \quad i \neq j$$

$$D_{ii} = I - T^*_{g_{i-1}^{-1} g_i} T_{g_{i-1}^{-1} g_i} \quad \text{for} \quad i = 2, 3, \ldots, n;$$

then $\widetilde{W} = (W_{ij})$, $\widetilde{D} = (D_{ij})$.

Since T_g are contractions, $\widetilde{D}$ is, clearly, a positive operator.

We now show that $\widetilde{T} = \widetilde{W}^* \widetilde{D} \widetilde{W}$.

As $\widetilde{T}$ and $\widetilde{W}^* \widetilde{D} \widetilde{W}$ are self-adjoint operators, it is sufficient to prove that $\widetilde{T}_{ij} = (\widetilde{W}^* \widetilde{D} \widetilde{W})_{ij}$ for $i \leqslant j$. At the same time $(\widetilde{W}^*)_{ij} = W^*_{ji} = 0$ for $j > i$, hence

$$(\widetilde{W}^* \widetilde{D} \widetilde{W})_{ij} = \sum_{k=1}^{i} (\widetilde{W}^*)_{ik} (\widetilde{D} \widetilde{W})_{kj}.$$

For $i = 1$ we have

$$(W^* D W)_{1j} = W^*_{11} D_{11} W_{1j} = W_{1j} = T_{g_1^{-1} g_j} = T_{1j},$$

and for $i > 1$

$$(\widetilde{W}^* \widetilde{D} \widetilde{W})_{ij} = \sum_{l=1}^{i} (\widetilde{W}^*)_{il} D_{ll} W_{lj} =$$

$$= \sum_{l=1}^{i} T_{g_l^{-1} g_i} D_{ll} T_{g_l^{-1} g_j} = T^*_{g_1^{-1} g_i} T_{g_1^{-1} g_j} +$$

$$+ \sum_{l=2}^{i} T^*_{g_l^{-1} g_i} T_{g_l^{-1} g_j} - \sum_{l=2}^{i} T^*_{g_l^{-1} g_i} T^*_{g_{l-1}^{-1} g} T_{g_{l-1}^{-1} g_l} T_{g_l^{-1} g_j}.$$

But from (9.1.3) and (9.1.4) there results

$$T^*_{g_l^{-1} g_i} T^*_{g_{l-1}^{-1} g_l} = T^*_{g_{l-1}^{-1} g_i}$$

$$T_{g_{l-1}^{-1} g_l} T_{g_l^{-1} g_j} = T_{g_{l-1}^{-1} g_j}$$

and therefore

$$(\widetilde{W}^*\widetilde{D}\widetilde{W})_{ij} = T^*_{g_1^{-1} g_i} T_{g_1^{-1} g_j} + \sum_{l=2}^{i} T^*_{g_l^{-1} g_i} T_{g_l^{-1} g_j} -$$

$$- \sum_{l=2}^{i} T^*_{g_{l-1}^{-1} g_i} T_{g_{l-1}^{-1} g_j} = T_{g_i^{-1} g_j} = T_{ij}.$$

This is nothing else than

$$\widetilde{T} = \widetilde{W}^*\widetilde{D}\widetilde{W}$$

which yields

$$(\widetilde{T}\widetilde{h}, \widetilde{h}) = (W^*\widetilde{D}\,\widetilde{W}\widetilde{h}, \widetilde{h}) = (\widetilde{D}\widetilde{W}\widetilde{h}, \widetilde{W}\widetilde{h}) \geqslant 0.$$

Thus, T is a positive operator, hence $g \to T_g$ satisfies (9.1.2). The theorem is completely proved. ◇

Let G, S, H be as above. *A semigroup of contractions* on H is a map $s \to T_s$ of S in $L(H)$, such that $T_1 = I$, $T_{s_1} T_{s_2} = T_{s_1 s_2}$, for any $s_1, s_2 \in S$ and $\|T_s\| \leqslant 1$ for any $s \in S$.

A unitary representation $g \to U_g$ of G on a Hilbert space K, is called a *unitary dilation* of the semigroup $\{T_s\}_{s\in S}$, if $H \subset K$ and

$$T_s h = PH_s h \qquad (s \in S, h \in H),$$

where P is the orthogonal projection of K on H.

Corollary 9.6. *If G is totally ordered by S, i.e. if $G = S \cup S^{-1}$, then any semigroup $\{T_s\}_{s\in S}$ of contractions on H has a unitary dilation.*

Proof. Indeed, if we write

$$T_g = \begin{cases} T_s & \text{if } g = s \in S \\ T_s^* & \text{if } g = s^{-1},\ s \in S \end{cases}$$

then we obtain a map $g \to T_g$ of G in $L(H)$ which, obviously, satisfies the conditions of Theorem 9.5. Therefore $g \to T_g$ verifies (9.1.2), hence admits a unitary dilation.

Theorem 9.7. *Let G, S, H be as written above, with $G = SS^{-1}$. Let $\{T_s\}_{s \in S}$ be a semigroup of contractions such that*

$$(9.1.5) \qquad T_{\sigma_1}^* T_{s_1} = T_{\sigma_2}^* T_{s_2} \quad \text{for} \quad s_1\sigma_1^{-1} = s_2\sigma_2^{-1} \notin S^{-1}$$

and

$$(9.1.6) \qquad T_\sigma^* T_s = T_s T_\sigma^* \quad \text{for} \quad s\sigma^{-1} \notin S \cup S^{-1}.$$

Then there exists a unitary dilation of the semigroup $\{T_s\}_{s \in S}$.

Proof. Let us write

$$T_g = \begin{cases} T_\sigma^* T_s & \text{if } g = s\sigma^{-1} \notin S^{-1} \\ T_\sigma^* & \text{if } g = \sigma^{-1},\ \sigma \in S. \end{cases}$$

According to (9.1.5), the mapping $g \to T_g$ is well defined and $T_1 = I$, $T_{g^{-1}} = T_g^*$, $\|T_g\| \leqslant 1$. We now show that it satisfies also (9.1.3). Let $g_1, g_2 \in G$ be such that g_1, g_2, $g_1g_2 \notin S^{-1}$. We put $g_1 = s_1\sigma_1^{-1}$, $g_2 = s\sigma_2^{-1}$. Since $g_1, g_2, g_1g_2 \notin S^{-1}$ we have

$$T_{g_1} = T_{\sigma_1}^* T_{s_1},\ T_{g_2} = T_{\sigma_2}^* T_{\sigma_2},\ T_{g_1g_2} = T_{\sigma_1\sigma_2}^* T_{s_1s_2}.$$

If g_1, $g_2 \in S$, then (9.1.3) is clearly satisfied. Suppose $g_2 \notin S$. Then

$$s_1s_2\sigma_2^{-1}s_1^{-1} = g_2 \notin S^{-1} \cup S$$

and, using (9.1.6), we obtain

$$T_{s_1 s_2} T^*_{\sigma_2 s_1} = T^*_{\sigma_2 s_1} T_{s_1 s_2}. \tag{9.1.7}$$

Now taking into account (9.1.7) we get

$$T_{g_1} T_{g_2} = T^*_{\sigma_1 \sigma_2} T^*_{\sigma_2 s_1} T_{s_1 s_2} = T^*_{\sigma_1 \sigma_2} T_{s_1 s_2} = T_{g_1 g_2}.$$

Similarly, we have

$$T_{g_2} T_{g_1} = T^*_{\sigma_2} T_{s_2} T^*_{\sigma_1} T_{s_1} = T^*_{\sigma_1 \sigma_2} T_{s_2 \sigma_1} T^*_{\sigma_1 \sigma_2} T_{s_1 \sigma_2} =$$

$$= T^*_{\sigma_1 \sigma_2} T^*_{\sigma_1 \sigma_2} T_{s_2 \sigma_1} T_{s_1 \sigma_2} = T^*_{\sigma_1 \sigma_2} T_{s_1 s_2} = T_{g_1 g_2}.$$

Therefore (9.1.3) is verified by the map $g \to T_g$. Now by Theorem 9.5, here exists a unitary dilation for the semigroup $\{T_s\}_{s \in S}$. ◇

Let G, S, H be as written above and $\{T_s\}_{s \in S}$ a semigroup of contractions on H.

A closed subspace M of H is called *invariant* (to $\{T_s\}_{s \in S}$) if $T_s M \subset M$ for any $s \in S$. The subspace M is said to be *doubly invariant* if $T_s M \subset M$, $T_s^* M \subset M$ for any $s \in S$.

If M is invariant, we denote by $T_s|M$ the restriction of T_s to M and by $\{T_s|M\}_{s \in S}$ the corresponding semigroup of contractions.

The semigroup of contractions $\{T_s\}_{s \in S}$ on H is called *unitary* if the operator T_s is unitary on H for any $s \in S$.

The semigroup $\{T_s\}_{s \in S}$ is called *completely non-unitary* if for any doubly invariant subspace M of H for which $\{T_s|M\}_{s \in S}$ is unitary we have $M = \{0\}$.

Theorem 9.8. *Let $\{T_s\}_{s \in S}$ be a semigroup of contractions on H. The space H has a unique decomposition under the form*

$$H = H_u \oplus H_c$$

such that H_u and H_c are doubly invariant, $\{T_s|H_u\}_{s \in S}$ is unitary and $\{T_s|H_c\}_{s \in S}$ is completely-non-unitary.

Proof. Let $\mathcal{M}$ be the family of doubly invariant subspaces M of H, with the property that $\{T_s|M\}_{s \in S}$ is unitary.

Let H_u be the subspace generated by $\bigcup_{M \in \mathcal{M}} M$. One easily verifies that H_u is doubly invariant and $\{T_s|H_u\}_{s \in S}$ is unitary. Hence H_u is the largest doubly invariant subspace M, for which $\{T_s|M\}_{s \in S}$ is unitary.

Let

$$H = H_u \oplus H_c.$$

H_c results doubly invariant. Let $M \subset H_c$ be double invariant, with $\{T_s|M\}_{s \in S}$ unitary. Then $M \subset H_u$ and, since it is orthogonal to H_u, we have $M = \{0\}$. Therefore $\{T_s|H_c\}_{s \in S}$ is a completely non-unitary semigroup.

Let

$$H = H_1 \oplus H_2$$

be another decomposition of H, with H_1, H_2 doubly invariant $\{T_s|H_1\}_{s \in S}$ unitary and $\{T_s|H_2\}_{s \in S}$ completely non-unitary. H_u being maximal there results $H_1 \subset H_u$.

Let us write

$$H_u = H_1 \oplus M.$$

Then M is doubly invariant and, as $M \subset H_u$, $\{T_s|M\}_{s \in S}$ results unitary. But M is orthogonal to H_1, hence $M \subset H_2$ and, as $\{T_s|H_2\}_{s \in S}$ is completely non-unitary, $M = \{0\}$. Thus $H_u = H_1$ and $H_s = H_2$, i.e. the decomposition is unique.

The theorem is therefore completely proved. ◇

Theorem 9.9. *Let $\{T_s\}_{s \in S}$ be a semigroup of contractions on H and $g \to U_g$ one of its unitary dilations.*

Let

$$H = H_u \oplus H_c$$

be the canonical decomposition of $\{T_s\}_{s \in S}$. We have

(9.1.8)
$$H_u = \{h \in H\colon U_g h \in H,\ h \in G\} =$$
$$= \{h \in H\colon \|T_s T_\sigma^* h\| = \|h\|,\ s, \sigma \in S\}.$$

Proof. Let

$$H_1 = \{h \in H: U_g h \in H,\ g \in G\}$$

$$H_2 = \{h \in H;\ \|T_s T_\sigma^* h\| = \|h\|,\ s, \sigma \in S\}.$$

For $h \in H_2$ we have

$$\|PU_s h\| = \|T_s h\| = \|h\| = \|U_s h\|$$

$$\|PU_{\sigma^{-1}} h\| = \|PU_\sigma^* h\| = \|T_\sigma^* h\| = \|h\| = \|U_{\sigma^{-1}} h\|.$$

Hence $U_s h \in H$, $U_{\sigma^{-1}} h \in H$ for any $s, \sigma \in S$. Then, for $g = s\sigma^{-1}$, we get

$$\|PU_g h\| = \|PU_s U_\sigma^* h\| = \|PU_s PU_\sigma^* h\| =$$

$$= \|T_s T_\sigma^* h\| = \|h\| = \|U_g h\|.$$

Therefore $U_g h \in H$ for any $g \in G$, which yields $H_2 \subset H_1$. H_1 is obviously a doubly invariant subspace and $\{T_s | H_1\}_{s \in S}$ is unitary. As H_u is maximal, there results $H_1 \subset H_u$. Since $T_s | H_u$ is unitary for any $s \in S$, there follows

$$\|T_s T_\sigma^* h\| = \|h\|$$

for any $h \in H_u$ and $s \in S$, hence $H_u \subset H_2$. Then

$$H_2 \subset H_1 \subset H_u \subset H_2$$

which proves relation (9.1.8). ◇

Remark. From (9.1.8) we obtain immediately

$$H_u = \{h \in H: \|T_\sigma^* T_s h\| = \|h\|,\ s, \sigma \in S\}.$$

9.2. The Wold decomposition

In the case $\{T_s\}_{s\in S}$ consists of isometries we obtain a decomposition theorem of the Wold type.

First we establish that any semigroup of isometries has a unitary dilation.

Theorem 9.10. (Ito) *Let G be an abelian group, $S \subset G$ a subsemigroup of G sucl that $G = SS^{-1}$, and H a Hilbert space. Every semigroup of isometries $\{T_s\}_{s\in S}$ has a unitary dilation.*

Proof. We write, for $g \in G$ of the form $g = \sigma^{-1} s$, $\sigma s \in S$,

$$T_g = T_\sigma^* T_s.$$

If $\sigma_1^{-1} s_1 = \sigma_2^{-1} s_2$, then $\sigma_2 s_1 = s_2 \sigma_1$ and therefore

$$T_{\sigma_2} T_{s_1} = T_{\sigma_1} T_{s_2}.$$

If we multiply at right by $T^*_{\sigma_1\sigma_2}$, we get

$$T^*_{\sigma_1} T_{s_1} = T^*_{\sigma_2} T_{s_2}.$$

Therefore the map $g \to T_g$ of G in $L(H)$ is well defined, and $T_1 = I$, $T_g^* = T_{g^{-1}}$. We now show that this map is positive definite. Let $g_1, \ldots, g_n \in G$, $g_i = \sigma_i^{-1} s_i$, σ_i, $s_i \in S$, $i = 1, 2, \ldots, n$, and $h_1, \ldots, h_n \in H$. Let $s = \sigma_1 \sigma_2 \ldots \sigma_n$ and $g_i' = s g_i$, $i = 1, 2, \ldots, n$. We obviously have $g_i' \in S$ for any i, and $g_i^{-1} g_j = g_i'^{-1} g_j'$ for any i, j. There results

$$T_{g_i^{-1} g_j} = T^*_{g_i'} T_{g_j'}$$

and then

$$\sum_{i,j} (T_{g_i^{-1} g_j} h_j, h_i) = \sum_{i,j} (T^*_{g_i'} T_{g_j'} h_j, h_i) =$$

$$= \| \sum_i T_{g_i'} h_i \|^2 \geqslant 0.$$

Thus $g \to T_g$ satisfies (9.1.2). Therefore, according to Theorem 9.1, $g \to T_g$ has a unitary dilation which, obviously, is a unitary dilation of the semigroup $\{T_s\}_{s \in S}$. ◇

The semigroup of isometries $\{T_s\}_{s \in S}$ will be called *quasiunitary* if

$$\text{(9.2.1)} \qquad \operatorname{clm}[\bigcup_{\sigma^{-1}s \notin S^{-1}} T_\sigma^* T_s H] = H.$$

The semigroup of isometries $\{T_s\}_{s \in S}$ is said *totally-non-unitary* if for any doubly invariant subspace M for which $\{T_s|M\}$ is quasiunitary, we have $M = \{0\}$.

It is clear that any unitary semigroup is quasiunitary, hence any totally non-unitary semigroup is completely non-unitary.

It also happens that a semigroup of isometries be quasiunitary and completely non-unitary at the same time. Such semigroups are called *singular*.

Let $\{T_s\}_{s \in S}$ be a quasiunitary semigroup and M a doubly invariant subspace. Then $\operatorname{clm}[\bigcup_{\sigma^{-1}s \notin S^{-1}} T_\sigma^* T_s M] \subset M$. Let $m \in M$ be orthogonal to $\operatorname{clm}[\bigcup_{\sigma^{-1}s \notin S^{-1}} T_\sigma^* T_s M]$ and $h \in H$. If $h = h_1 + h_2$, with $h_1 \in M$ and $h_2 \in M^\perp$, we have

$$(m, T_\sigma^* T_s h) = (m, T_\sigma^* T_s h_1) + (m, T_\sigma^* T_s h_2) = 0$$

for $\sigma^{-1} s \notin S^{-1}$. Therefore m is orthogonal to $\operatorname{clm}[\bigcup_{\sigma^{-1}s \notin S^{-1}} T_\sigma^* T_s H] = H$, i.e. $m = 0$. Hence

$$M = \operatorname{clm}[\bigcup_{\sigma^{-1}s \notin S^{-1}} T_\sigma^* T_s M],$$

which means $\{T_s|M\}_{s \in S}$ quasiunitary.

If M is a doubly invariant subspace, then $\{T_s|M\}_{s \in S}$ *is unitary (quasiunitary, completely nonunitary, totally nonunitary, singular)* if $\{T_s\}_{s \in S}$ *has the same property.*

Theorem 9.11. *Let* $\{T_s\}_{s \in S}$ *be a semigroup of isometries on* H. *The space* H *admits a unique decomposition under the form*

$$H = H_q \oplus H_t$$

where H_q, H_t are doubly invariant subspaces, $\{T_s|H_q\}_{s\in S}$ is quasiunitary and $\{T_s|H_t\}_{s\in S}$ is totally nonunitary.

Proof. We put

(9.2.2)
$$N = [\bigcup_{s\sigma^{-1}\notin S^{-1}} T_\sigma^* T_s H]^\perp.$$

For $n \in N$, $h \in H$ and $\sigma^{-1} s \notin S$, we have

$$(T_\sigma^* T_s n, h) = (n, T_s^* T_\sigma h) = 0.$$

hence

(9.2.3)
$$T_\sigma^* T_s n = 0 \qquad (n \in N, \ \sigma^{-1} s \notin S).$$

The subspace $T_s N$ are pairwise orthogonal. Indeed, since $S \cap S^{-1} = \{1\}$ we get

$$(T_s n, T_\sigma m) = (n, T_s^* T_\sigma m) = (T_\sigma^* T_s n, m) = 0$$

for any $m, n \in N$, $s, \sigma \in S$, $s \neq \sigma$.

Let us write

(9.2.4)
$$H_t = \bigoplus_{s\in S} T_s N$$

and

$$H = H_q \oplus H_t.$$

H_q and H_t are doubly invariant subspaces. Indeed, H_t is obviously invariant. If $m \in H_q$ and $s \in S$, then from (9.2.3)

$$(T_\sigma m, T_s n) = (m, T_\sigma^* T_s n) = 0$$

for any $\sigma \in S$ for which $\sigma^{-1} s \notin S$ and for any $n \in N$. If $\sigma^{-1} s = s_1 \in S$, then $s = \sigma s_1$ and

$$(T_\sigma m, T_s n) = (m, T_\sigma^* T_\sigma T_{s_1} n) = (m, T_{s_1} n) = 0,$$

for any $n \in N$, following the definition of H_q. Hence $T_\sigma m$ is orthogonal to H_t, that is $T_\sigma m \in H_q$. Therefore H_q is also invariant, which implies that H_t and H_q are doubly invariant.

The semigroup $\{T_s|H_q\}_{s\in S}$ is quasiunitary. Indeed, since H_q is doubly invariant we have:

$$\text{clm}\,[\bigcup_{\sigma^{-1}s\notin S^{-1}} T_\sigma^* T_s H_q] \subset H_q.$$

Let $m \in H_q$ be orthogonal to $[\bigcup_{\sigma^{-1}s\notin S^{-1}} T_\sigma^* T_s H_q]$ and $h \in H$ of the form $h = h_1 + h_2$, $h_1 \in H_q$, $h_2 \in H_t$.

Then

$$(m, T_\sigma^* T_s h) = (m, T_\sigma^* T_s h_1) + (m, T_\sigma^* T_s h_2) = 0$$

for any $s, \sigma \in S$, $\sigma^{-1}s \notin S^{-1}$. Hence m is orthogonal to $\text{clm}\,[\bigcup_{\sigma^{-1}s\notin S^{-1}} T_\sigma^* T_s H] \supset H_q$ and therefore $m = 0$. There results

$$H_q = \text{clm}\,[\bigcup_{s\sigma^{-1}\notin S^{-1}} T_\sigma^* T_s H_q]$$

i. e. $\{T_s|H_q\}_{s\in S}$ is quasiunitary.

Let now M be a doubly invariant subspace included in H_t, such that $\{T_s|M\}_{s\in S}$ is quasiunitary. Since $M = \text{clm}\,[\bigcup_{\sigma^{-1}s\notin S^{-1}} T_\sigma^* T_s M]$, M results orthogonal to N. We then have

$$(m, T_s n) = (T_s^* m, n) = 0$$

for any $m \in M$, $n \in N$ and $s \in S$. Hence M is orthogonal to $H_t \supset M$, i.e. $M = \{0\}$. Therefore $\{T_s|H_t\}_{s\in S}$ is totally nonunitary.

Now let

$$H = H_1 \oplus H_2$$

be a new decomposition of H, where H_1, H_2 are doubly invariant, $\{T_s|H_1\}_{s\in S}$ is quasiunitary and $\{T_s|H_2\}_{s\in S}$ is totally nonunitary. As H_1

is doubly invariant and $H_1 = \text{clm}\,[\bigcup_{\sigma^{-1}s \notin S^{-1}} T_\sigma^* T_s H_1]$ it is orthogonal to H_t (as above), that is $H_1 \subset H_q$. Write

$$H = H_1 \oplus M.$$

Then M is clearly doubly invariant and, since $M \subset H_q$, $\{T_s|M\}_{s\in S}$ is quasiunitary. But M is orthogonal to H, hence $M \subset H_2$ and since, $\{T_s|H_2\}_{s\in S}$ is totally nonunitary, we have $M = \{0\}$. Therefore $H_q = H_1$, $H_t = H_2$.

The proof is complete. ◇

Now, if we combine Theorem 9.8 with Theorem 9.10 we obtain the following generalization of the Wold decomposition theorem.

Theorem 9.12. *Let $\{T_s\}_{s\in S}$ be a semigroup of isometries on H. The space H has a unique decomposition of the form*

$$H = H_u \oplus H_\sigma \oplus H_t$$

where H_u, H_σ, H_t are doubly invariant subspaces, $\{T_s|H_u\}_{s\in S}$ is unitary, $\{T_s|H_\sigma\}_{s\in S}$ is singular and $\{T_s|H_t\}_{s\in S}$ is totally nonunitary.

Proof. Applying successively Theorem 9.8 and Theorem 9.10 we get the decomposition. To prove the uniqueness it is sufficient to observe, for instance, that $H_\sigma \oplus H_t$ is the completely nonunitary part of $\{T_s\}_{s\in S}$.

9.3. The semigroup of unilateral translations

Let N be a Hilbert space; we denote by $L^2(N, G)$ the Hilbert space of families $\tilde{h} = (h_g)$, $g \in G$, $h_g \in N$, with the property

$$\|\tilde{h}\| \leqslant \sum_{g\in G} \|h_g\|^2 < \infty,$$

with the usual operations and the inner product

$$(\tilde{h}, \tilde{k}) = \sum_{g\in G} (h_g, k_g),$$

$$\tilde{h} = (h_g),\ \tilde{k} = (k_g),$$

$H^2(N; S)$ will denote the closed subspace of $L^2(N; G)$ consisting of elements $\widetilde{h} = (h_g) \in L^2(N; G)$ with $h_g = 0$ for $g \notin S$.

If P is the orthogonal projection of $L^2(N; G)$ on $H^2(N; S)$, then $P(h_g) = (h_g')$, where

$$h_g' = \begin{cases} h_g & \text{if } g \in S \\ 0 & \text{if } g \notin S. \end{cases}$$

The map $n \to (n_g)$ defined on N with values in $H^2(N; S)$ by $n_g = 0$ if $g \neq 1$ and $n_1 = n$, is an isometric embedding of N in $H^2(N; S)$.

For $s \in S$ we defined on $H^2(N; S)$ the operator

$$T_s(h_g) = (h_{s^{-1}g}). \tag{9.3.1}$$

As we can easily verify, $\{T_s\}_{s \in S}$ is a semigroup of isometries on $H^2(N; S)$. We call this semigroup, the *semigroup of unilateral translations in* $H^2(N; S)$.

If $\{T_s\}_{s \in S}$ is the semigroup of unilateral translations in $H^2(N; S)$ then for any $s \in S$, we have

$$T_s^*(h_g) = P(h_{sg}) \tag{9.3.2}$$

and

$$H^2(N; S) = \bigoplus_{s \in S} T_s N. \tag{9.3.3}$$

Theorem 9.13. *Let* $\{T_s\}_{s \in S}$ *be a semigroup of isometries on* H. *The following assertions are equivalent.*

a) *There exists a Hilbert space* N *such that the semigroup* $\{T_s\}_{s \in S}$ *is unitary equivalent to the semigroup of unilateral translations in* $H^2(N; S)$.

b) *The semigroup* $\{T_s\}_{s \in S}$ *is totally nonunitary.*

c) *If* M *is a doubly invariant subspace of* H, *with* $M \subset$

$$\subset \operatorname{clm}[\bigcup_{\sigma^{-1}s \notin S^{-1}} T_\sigma^* T_s H], \textit{ then } M = \{0\}.$$

d) *If* M *is a doubly invariant subspace of* H, *with*

$$M \cap [\bigcup_{\sigma^{-1}s \notin S^{-1}} T_\sigma^* T_s H]^\perp = \{0\}, \textit{ then } M = \{0\}.$$

Proof. The implications d) → c), c) → b) are obvious.

b) → a). Let $\{T_s\}_{s\in S}$ be totally nonunitary. Then, from Theorem 3.5, there results

$$H = H_t = \bigoplus_{s\in S} T_s N \tag{9.3.4}$$

where

$$N = [\bigcup_{\sigma^{-1}s\notin S^{-1}} T_\sigma^* T_s H]^\perp.$$

The mapping $h \to (h_g)$, with $h_s = n_s$ if $h = \sum_{s\in S} T_s n_s$ and $h_g = 0$ for $g \notin S$, is a unitary equivalence between H and H^2 $(N; S)$; it transforms the semigroup $\{T_s\}_{s\in S}$ in the semigroup of unilateral translations in $H^2(N; S)$. Therefore b) → a) is proved.

a) → c). Let N be a Hilbert space and $\{T_s\}_{s\in S}$ the semigroup of unilateral translations in $H^2(N; S)$. We first prove the relation

$$N = [\bigcup_{\sigma^{-1}s\notin S^{-1}} T_\sigma^* T_s H^2(N; S)]^\perp \tag{9.3.5}$$

where N is embedded in H^2 $(N; S)$ as seen before.

Let $\tilde{n} \in N$, $\tilde{n} = (n_g)$, $n_g = 0$ for $g \neq 1$. Using (9.3.2) we have

$$T_\sigma^* T_s \tilde{n} = T_\sigma^* T_s(n_g) = T_\sigma^*(n_{s^{-1}g}) = P(n_{s^{-1}\sigma g}) = (n'_g)$$

where

$$n'_g = \begin{cases} n_{s^{-1}\sigma g} & \text{if } g \in S \\ 0 & \text{if } g \notin S. \end{cases}$$

Hence $n_g = 0$ for $g \neq 1$ and therefore $n'_g = 0$ for $g \neq \sigma^{-1}s$. Then, for $n \in N$ and $\sigma^{-1} s \notin S$ we get

$$T_\sigma^* T_s \tilde{n} = 0. \tag{9.3.6}$$

This yields

$$(\tilde{n}, T_s^* T_\sigma \tilde{h}) = (T_\sigma^* T_s \tilde{n}, \tilde{h}) = 0$$

for $s, \sigma \in S$, $s\sigma^{-1} \notin S$. Hence

$$N \subset [\bigcup_{\sigma^{-1}s \notin S^{-1}} T_\sigma^* T_s H^2(n; S)]^\perp.$$

Conversely, if $\tilde{n} = (n_g) \in [\bigcup_{\sigma^{-1}s \notin S^{-1}} T_\sigma^* T_s H^2(N; S)]^\perp$ then we have

$$(T_\sigma^* T_s \tilde{n}, \tilde{h}) = (\tilde{n}, T_s^* T_\sigma \tilde{h}) = 0,$$

for any $\tilde{h} \in H^2(N; S)$, $\sigma^{-1} s \notin S$. Therefore

$$T_\sigma^* T_s \tilde{n} = 0$$

for any $s, \sigma \in S$, $s\sigma^{-1} \notin S$. In particular $T_\sigma^* \tilde{n} = 0$ for any $\sigma \in S$, $\sigma \neq 1$. If $T_\sigma^* \tilde{n} = (n_g')$, then $n_g' = 0$ for any $g \in G$. But $n_\sigma = n_1' = 0$; hence $n_\sigma = 0$ for any $\sigma \neq 1$, which yields $n \in N$. The relation (9.3.5) is proved.

Let now M be a doubly invariant subspace included in $\text{clm}\,[\bigcup_{\sigma^{-1}s \notin S^{-1}} T_\sigma^* T_s H^2(N; S)]$. From (9.3.5) there results $M \subset N^\perp$ and, since M is doubly invariant, we obtain

$$(T_s n, m) = (n, T_s^* m) = 0.$$

Therefore M is orthogonal to $T_s N$ for any $s \in S$ and, from (9.3.2), there follows $M = \{0\}$.

b) → d). Let M be a doubly invariant subspace of H, such that

$$M \cap [\bigcup_{\sigma^{-1}s \notin S^{-1}} T_\sigma^* T_s H]^\perp = \{0\}. \tag{9.3.7}$$

We have

$$M = \text{clm}\,[\bigcup_{\sigma^{-1}s \notin S^{-1}} T_\sigma^* T_s M]. \tag{9.3.8}$$

Indeed, since M is doubly invariant, there results

$$\text{clm}\,[\bigcup_{\sigma^{-1}s \notin S^{-1}} T_\sigma^* T_s M] \subset M.$$

Let $m \in M$ be orthogonal to $[\bigcup_{\sigma^{-1}s \notin S^{-1}} T_\sigma^* T_s M]$, and $h \in H$ be of the form $h = h_1 + h_2$, with $h_1 \in M$ and $h_2 \in M^\perp$. Then

$$(m, T_\sigma^* T_s h) = (m, T_\sigma^* T_s h_1) + (m, T_\sigma^* T_s h_2) = 0$$

for any $s, \sigma \in S, \sigma^{-1}s \notin S^{-1}$. Hence $m \in [\bigcup_{\sigma^{-1}s \notin S^{-1}} T^* T_s H]^\perp$, and, from (9.3.7), there follows $m = 0$. Therefore (9.3.8) holds.

From (b) there results $M = \{0\}$ and b) → d) is proved.

Since the following implications are true

$$\begin{array}{ccc} a & \longleftarrow & b \\ \downarrow & \nearrow & \downarrow \\ c & \longleftarrow & d \end{array}$$

the theorem is proved. ◇

The following theorem is a completion to Theorem 9.13 in the case N is one-dimensional.

Theorem 9.14. *Let $\{T_s\}_{s \in S}$ be a semigroup of isometries on H. The semigroup $\{T_s\}_{s \in S}$ is unitary equivalent to the semigroup of unilateral translations on a $H^2(N; S)$, with N one dimensional, if and only if $\{T_s\}_{s \in S}$ is not quasiunitary and has no proper doubly invariant subspaces.*

Proof. Let $\{T_s\}_{s \in S}$ be the semigroup of unilateral translations on $H^2(N; S)$, with N one-dimensional. Let $M \subset H^2(N; S)$ be a doubly invariant subspace of $H^2(N; S)$. If

$$M \cap [\bigcup_{\sigma^{-1}s \notin S^{-1}} T_\sigma^* T_s H^2(N; S)]^\perp = 0$$

then from Theorem 3.7 there follows $M = \{0\}$.

Suppose

$$M \cap [\bigcup_{\sigma^{-1}s \notin S^{-1}} T_\sigma^* T_s H^2(N; S)]^\perp \neq 0.$$

From (3.4.5) there results $M \cap N \neq \{0\}$. As N is one-dimensional, we have $M \cap N = N$, therefore $H^2(N; S) = \bigoplus_{s \in S} T_s N \subset M$.

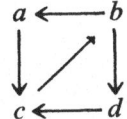

Let now $\{T_s\}_{s \in S}$ be a semigroup of isometries on H which is not quasiunitary and has no proper doubly invariant subspaces. Then, clearly $\{T_s\}_{s \in S}$ satisfies point d) of Theorem 9.13. Hence $\{T_s\}_{s \in S}$ is unitarily equivalent to the semigroup of unilateral translations in an $H^2(N; S)$. Let $n_0 \in N$, $n_0 \neq 0$ and let N_0 be the one-dimensional space spanned by n_0. Let us write

$$M = \bigoplus_{s \in S} T_s N_0.$$

M is a doubly invariant space. Indeed, if $m = \sum_{s \in S} T_s n_s$ with $n_s \in N_0$, then

$$T_\sigma m = \sum_{s \in S} T_{s\sigma} n_s \in M$$

$$T_\sigma^* m = \sum_{s_1 = \sigma^{-1} s} T_{s_1} n_s \in M.$$

Since $M \cap N = N_0 \neq 0$, we have $H = M$, which implies that $N = N_0$ is one-dimensional.

The theorem is proved. ◇

9.4. Representations of S-generated algebras

Let A be an S-generated algebra on X such that $S \cap S^{-1} = \{1\}$, H a Hilbert space and $f \to T_f$ a representation of A on H. Then it is clear that $\{T_s\}_{s \in S}$ is a semigroup of contractions on H.

Moreover, for any finite systems $s_1, \ldots, s_n \in S$ and $c_1, \ldots, c_n \in C$, we have

$$\| \Sigma\, c_i T_{s_i} \| \leqslant \| \Sigma\, c_i s_i \| \tag{9.4.1}$$

hence the representation $f \to T_f$ is generated by the semigroup $\{T\}_{s \in S}$. Conversely, if $\{T_s\}_{s \in S}$ is a semigroup of contractions on H, which satisfies (9.4.1) then, if we write

$$T_{\Sigma c_i s_i} = \Sigma\, c_i T_{s_i}$$

and extend by continuity, we clearly obtain a representation $f \to T_f$ on A on H.

Theorem 9.15. (i) *The subspace M of H is spectral for $f \to T_f$ if and only if M is doubly invariant to $\{T_s\}_{s\in S}$ and the semigroup $\{T_s|M\}_{s\in S}$ is unitary.*

(ii) *The representation $f \to T_f$ is spectral if and only if $\{T_s\}_{s\in S}$ is unitary.*

(iii) *The representation $f \to T_f$ is completely non-spectral if and only if $\{T_s\}_{s\in S}$ is completely non-unitary.*

(iv) *The representation $f \to T_f$ and the semigroup $\{T_s\}_{s\in S}$ have the same canonical decomposition.*

Proof. (i) Supose M is spectral for $f \to T_f$ and let $\varphi \to U_\varphi$ be the extension to $C(X)$ of the representation $f \to T_f|M$. Then, clearly, M is doubly invariant to $\{T_s\}_{s\in S}$ and

$$T_s^* T_s m = U_s^* U_s m = U_{\bar{s}s} m = m = U_s U_{\bar{s}} m = T_s T_s^* m$$

for any $m \in M$ and $s \in S$. Therefore $\{T_s|M\}_{s\in S}$ is unitary.

Assume now that M is doubly invariant to $\{T_s\}_{s\in S}$ and $\{T_s|M\}_{s\in S}$ is unitary. For $f \in A$ of form $f = \Sigma c_i s_i$ we have $T_f M \subset M$, $T_f^* M \subset M$ and by continuity, M results doubly invariant to $f \to T_f$.

For a function $\varphi \in C(X)$ of the form $\varphi = \Sigma c_i s_i \bar{s}$ and $m \in M$, we write

$$U_\varphi m = \Sigma\, c_i T_{s_i} T_s^* m.$$

If $\varphi_1 = \Sigma\, a_i s_i \bar{s}$, $\varphi_2 = \Sigma\, b_i \sigma_i \bar{\sigma}$, then

$$\varphi_1 + \varphi_2 = \Sigma\, (a_i s_i \sigma + b_i \sigma_2 s)\, \overline{s\sigma}$$

hence

$$U_{\varphi_1+\varphi_2} m = \Sigma\, (a_i T_{s_i} T_\sigma + b_i T_{\sigma_i} T_s)\, T_s^* T_\sigma^* m =$$

$$= \Sigma\, a_i T_{s_i} T_s^* m + \Sigma\, b_i T_{\sigma_i} T_\sigma^* m = U_{\varphi_1} m + U_{\varphi_2} m$$

for any $m \in M$. We used here the fact that $T_s|M$ is unitary for any $s \in S$.

Moreover

$$\|U_\varphi m\| = \|\Sigma\, c_i T_{s_i} T_s^* m\| \leqslant \|\Sigma\, c_i T_{s_i}\|\, \|T_s m\| = \|\Sigma\, c_i T_{s_i}\|\, \|m\| \leqslant$$

$$\leqslant \|\Sigma\, c_i s_i\, \|m\| = \|\varphi\|\, \|m\|.$$

Therefore the map $\varphi \to U_\varphi$ of $C(X)$ in $L(M)$, which is well defined on elements of form $\varphi = \Sigma c_i s\bar{s}$, is linear and continuous. We can easily verify that $\varphi \to U_\varphi$ is also multiplicative. Since the elements φ of the form $\varphi = \Sigma c_i s_i \bar{s}$ are dense in $C(X)$, $\varphi \to U_\varphi$ may be extended to a representation of $C(X)$ on M which, obviously, is an extension to $C(X)$ of the representation $f \to T_f|M$. Hence, the subspace M is spectral for $f \to T_f$.

(ii), (iii) an (iv) are immediate consequences of point (i). ◇

We shall further assume that S is analytic free.

According to Theorem 6.18 we are, in fact, in the following context: X is the dual of the abelian discrete group $G = SS^{-1}$ and A is the algebra $A(S)$-generated in $C(X)$ by the elements of S, viewed as characters on X.

Theorem 9.16. *Let A be an S-generated algebra on X and $f \to T_f$ a representation of A on H. The representation $f \to T_f$ has a spectral dilation if and only if the semigroup $\{T_s\}_{s \in S}$ has a unitary dilation.*

Proof. Let $\varphi \to U_\varphi$ be a spectral dilation of $f \to T_f$. Then the map $g \to U_g$ is a unitary dilation of the semigroup $\{T_s\}_{s \in S}$.

Let now $g \to U_g$ be a unitary dilation of the semigroup $\{T_s\}_{s \in S}$. From von Neumann formula (Corollary 9.3) there results

$$\|\Sigma c_i U_{g_i}\| \leqslant \|\Sigma c_i g_i\|$$

for any finite system $c_1,\ldots,c_n$ of constants and any finite system $g_1,\ldots, g_n \in G$. For φ of the form $\varphi = \Sigma c_i s_i \bar{s}$, s_i, $s \in S$, we may define

$$U_\varphi = \Sigma c_i U_{s_i} U_s^*$$

and, using the same argument as in Theorem 9.15, we obtain the representation $\varphi \to U_\varphi$ of $C(X)$ on K (the dilation space), which is a spectral dilation of the representation $f \to T_f$.

Theorem 9.17. *The representation $f \to T_f$ of the S-generated algebra A is subspectral if and only if the semigroup $\{T_s\}_{s \in S}$ is isometric.*

Proof. Assume $f \to T_f$ is subspectral and let $\varphi \to U_\varphi$ be its spectral dilation for which $U_f h \in H$ for $f \in A$ and $h \in H$. We have

$$\|T_s h\| = \|U_s h\| = \|h\|$$

for any $s \in S$ and $h \in H$. Hence T_s is an isometry on H for any $s \in S$.

Conversely, assume T_s isometric. Any semigroup of isometries admits a unitary dilation (Theorem 9.10). Let $\{U_g\}_{g\in S}$ be the unitary dilation of $\{T_s\}_{s\in S}$. Then

$$\|PU_s h\| = \|T_s h\| = \|h\| = \|U_s h\|$$

hence $U_s h \in H$ for any $s \in S$ and $h \in H$. Let $\varphi \to U_\varphi$ be the spectral dilation of $f \to T_f$, constructed as in Theorem 9.16. For $f \in A$, $f = \Sigma c_i s_i$, we have

$$U_f h = \sum c_i U_{s_i} h \in H \qquad (h \in H)$$

and, therefore, it is clear that $U_f h \in H$ for any $f \in A$ and $h \in H$. Thus the representation $f \to T_f$ is subspectral.

Let m be the Haar measure of X. We say that the representation $f \to T_f$ of A on H is *m-spectral* (*m-completely non-spectral, m-singular*) if the semigroup of contractions $\{T_s\}_{s\in S}$ is quasiunitary (totally non-unitary, singular). If $G = S \cup S^{-1}$ then, m has the uniqueness property and, as one easily verifies, these definitions are in agreement with those given in paragraph 8.4.

Now, Theorem 9.11 may be transposed in a function language.

Theorem 9.18. *Let $f \to T_f$ be a subspectral representation of the S-generated algebra A on H. The space H has a unique decomposition under the form*

$$H = H_s \oplus H_\sigma \oplus H_t$$

where H_s, H_σ, H_t are doubly invariant subspace, $f \to T_f|H_s$ is a spectral representation, $f \to T_f|H_\sigma$ is m-singular and $f \to T_f|H_t$ is m-completely non-singular.

In the same sense we write again Theorems 9.13 and 9.14.

Theorem 9.19. *The subspectral representation $f \to T_f$ of the S-generated algebra A on H is unitarily equivalent to the natural representation of A on $H^2(N; \mathrm{d}m)$ if and only if it is m-completely non-spectral.*

The representation $f \to T_f$ is unitary equivalent to the natural representation of A on $H^2(\mathrm{d}m) = H^2(N; \mathrm{d}m)$, with N one-dimensional, if and only if $f \to T_f$ is not m-spectral and has no proper doubly invariant subspaces.

9.5. Prediction theorems

Let A be a function algebra on X and m a representing measure for A. All along this paragraph A and m are supposed to verify one of the following assertions: (1) m is a representing measure for A with the uniqueness property. (2) A is an S-generated algebra, with S analytic free and m is the Haar measure of X, viewed as the dual group of the discrete abelian group $G = SS^{-1}$.

Moreover, if in (2) we assume $G = S \bigcup S^{-1}$ then, clearly, for A and m satisfying (2), the assertion (1) is also true. But, as we have seen (in the case of the H^∞ algebra), A and m may satisfy (1) without satisfying (2), hence the two cases do not include each other.

In these two cases we enounced and proved a decomposition of Wold type for the subspectral representations. In the present paragraph we shall analyse the Wold decomposition for the natural representations of A on $H^2(\mathrm{d}\mu)$, where μ is a positive measure on X.

The notions of m-spectral, m-completely nonspectral and m-singular representation will have the sense required by the situation.

Proposition 9.20. *Suppose that A and m satisfy (1) (respectively) (2) and let* μ *be a positive measure on X with* $\mu(X) = 1$. *If* $N = [\bigcup_{f \in A_m} T_f H^2(\mathrm{d}\mu)]^\perp$ *(respectively* $N = [\bigcup_{\sigma^{-1}s \notin S^{-1}} T_s^* T_s H^2(\mathrm{d}\mu)]^\perp$) *then* $\dim N \leqslant 1$.

Proof. In both situations one easily verifies that $C + N^\perp$ is dense in $H^2(\mathrm{d}\mu)$. Let P be the projection of $H^2(\mathrm{d}\mu)$ on C. We have $P(n_1 - n_2) = 0$ for $n_1, n_2 \in N$, with $Pn_1 = Pn_2$, hence

$$(n_1 - n_2,\ c + n^\perp) = (n_1 - n_2,\ c) + (n_1 - n_2,\ n^\perp) = 0$$

for any $c \in C$ and $n^\perp \in N^\perp$. Then $n_1 = n_2$ and therefore $P|N$ is an injection of N in C, that is $\dim N \leqslant 1$. ◇

Corollary 9.21. *The natural representation* $f \to T_f$ *of A on* $H^2(\mathrm{d}\mu)$ *is m-completely nonspectral if and only if it is not m-spectral and has no proper doubly invariant subspaces.*

Proof. The proof follows from Theorem 9.19 and the preceding proposition.

Proposition 9.22. *Any doubly invariant subspace* $M \neq 0$ *of the natural representation* $f \to T_f$ *of A on* $H^2(\mathrm{d}\mu)$ *has the form* $M = \chi_E L^2(\mathrm{d}\mu) \bigcap H^2(\mathrm{d}\mu)$, *where E is a Borel subset of X. E is* μ*-essentially unique determined by M.*

Proof. Let $L(M)$ be the minimal subspace of $L^2(\mathrm{d}\mu)$, invariant to the multiplication by functions of $C(X)$ and which contains M. The elements $\bar{g}m$ with $g \in A$ and $m \in M$ span $L(M)$. If P is the projection of $L^2(\mathrm{d}\mu)$ on $H^2(\mathrm{d}\mu)$, we have

$$P\bar{g}m = T_g^* m \in M.$$

Hence

$$PL(M) \subset M \subset PL(M),$$

that is

$$M = PL(M).$$

Since $L(M)$ is an invariant subspace to the multiplication by functions of $C(X)$, it has the form

$$L(M) = \chi_E L^2(\mathrm{d}\mu)$$

where E is a Borel set, μ-essentially unique determined (Theorem7.14). There results

$$\chi_E L^2(d\mu) \cap H^2(\mathrm{d}\mu) \subset PL(M) = M \subset \chi_E L^2(\mathrm{d}\mu) \cap H^2(\mathrm{d}\mu)$$

which yields

$$M = \chi_E L^2(\mathrm{d}\mu) \cap H^2(\mathrm{d}\mu).$$

The proposition is therefore proved. ◇

A positive measure μ on X, $\mu(X) = 1$, will be called a *Szegö-total measure* if it is not a point measure and if for any Borel subset E of X, with $\chi_E\, L^2(\mathrm{d}\mu) \cap H^2(\mathrm{d}\mu) \neq \{0\}$ we have $\mu(E) = 1$.

There immediately results that *the non-point measure μ is Szegö-total if and only if any function of $H^2(\mathrm{d}\mu)$, which vanishes on a set of positive measure, is equal to zero almost everywhere on X.*

Proposition 9.23. *Any Szegö-total measure on X is a Szegö measure.*

Proof. Suppose that a Szegö-total measure is not a Szegö measure. Then there exists a Borel subset E of X, with $\mu(E) > 0$, such that $\chi_E L^2(\mathrm{d}\mu) \subset H^2(\mathrm{d}\mu)$. Therefore $\chi_E L^2(\mathrm{d}\mu) \cap H^2(\mathrm{d}\mu) = \chi_E L^2(\mathrm{d}\mu) \neq 0$. Since

μ is a Szegö-total measure, we have $\mu(E) = 1$, hence $L^2(d\mu) = \chi_E L(d\mu) \subset H^2(d\mu) \subset L^2(d\mu)$, that is $L^2(d\mu) = H^2(d\mu)$.

This yields $\chi_E L^2(d\mu) \subset H^2(d\mu)$ for any Borel set E, hence $\mu(E) = 0$ or $\mu(E) = 1$ for any Borel set E of X; that means μ is a point measure which contradicts the fact that μ is a Szegö-total measure. ◇

Theorem 9.24. *Let μ be a positive measure on X, $\mu(X) = 1$, and $f \to T_f$ be the natural representation of A on $H^2(d\mu)$. Then*

(i) *If μ is a Szegö-singular measure, $f \to T_f$ is a spectral representation.*

(ii) *If μ is a Szegö measure and $N = [\bigcup_{\sigma^{-1}s \notin S^{-1}} T_\sigma^* T_s H^2(d\mu)]^\perp = 0, f \to T_f$ is m-singular.*

(iii) *If μ is a Szegö-total measure and $N \neq 0$, $f \to T_f$ is m-completely nonspectral.*

Proof. Points (i) and (ii) result from Theorem 8.19. Point (iii) follows from Corollary 9.17 and Proposition 9.18.

We know that, for (i) and (ii) the converse assertions are also true. As concerns point (iii), we have the following statement:

Proposition 9.25. *If $f \to T_f$ is m-completely nonspectral, then for any Borel set E of X, with $\chi_E \in H^2(d\mu)$, we have $\mu(E) = 0$ or $\mu(E) = 1$.*

Proof. Let E be a Borel set with $\chi_E \in H^2(d\mu)$. Then $M = \chi_E H^2(d\mu)$ is a doubly invariant subspace of $H^2(d\mu)$. Indeed, $T_g M \subset M$ for any $g \in A$. Let $h \in M$ and $gh = h_1 + h_2$ with $h_1 \in M$ and $h_2 \in \chi_E\, L^2(d\mu)$ orthogonal to M. We have

$$\int k\bar{h}_2 d\mu = \int \chi_E k\bar{h}_2\, d\mu = 0$$

for any $k \in H^2(d\mu)$, hence $h_2 \in H^2(d\mu)^\perp$. There results $T_g^* h = P\bar{g}h = h_1 \in M$, and therefore $T_g^* M \subset M$. Thus M is a doubly invariant subspace and, since $f \to T_f$ is totally nonspectral, from Corollary 9.17 we get $M = \{0\}$ or $M = H^2(d\mu)$, i.e. $\mu(E) = 0$ or $\mu(E) = 1$. ◇

We now show that Theorem 9.24 may be interpreted as a prediction theorem.

Let K be a Hilbert space (K may be, for instance, the space $L^2(dP)$ of square integrable functions on a probability space), G an abelian group and $S \subset G$ a subsemigroup of G such that $G = S \cup S^{-1}$, $S \cap S^{-1} = \{1\}$.

We say that the map $g \to F_g$ of G in K is a G-stationary process, if

$$(F_{gg_1}, F_{gg_2}) = (F_{g_1}, F_{g_2}) \qquad (g, g_1, g_2 \in G). \tag{9.5.1}$$

The prediction problem relative to the future S, for such a process, is that of approximating an element F_g by linear combinations of elements F_γ with $g^{-1}\bar{\gamma} \notin S$.

More precisely, let us suppose, without any loss in generality, that the space $\mathscr{H}$ is the closure of the set spanned by the elements F_g, $g \in G$. For any $g \in G$ we put

$$\mathscr{M}_g = \text{clm}\,[\{F_\gamma;\quad \gamma \in G,\quad g^{-1}\gamma \notin S\}].$$

If $g_1 \leqslant g_2$, i.e. if $g_2 g_1^{-1} \in S$, then $g_1^{-1}\gamma \notin S$ yields $g_2^{-1}\gamma \notin S$, hence $\mathscr{M}_{g_1} \subset \mathscr{M}_{g_2}$. Thus, $\{\mathscr{M}_g\}$ is an increasing family of subspaces of $\mathscr{H}$. Let us write

$$\mathscr{H}_1 = \bigcap_{g \in G} \mathscr{M}_g.$$

Let E_g be the orthogonal projection of F_g on the orthogonal complement of $\mathscr{M}_g$. For any $g \in G$, E_g is orthogonal to $\mathscr{H}_1$ and $\{E_g;\ g \in G\}$ is an orthogonal set. It spans the closed subspace $\mathscr{H}_3$ of $\mathscr{H}$, orthogonal to $\mathscr{H}_1$. We put

$$\mathscr{H} = \mathscr{H}_1 \oplus \mathscr{H}_2 \oplus \mathscr{H}_3. \tag{9.5.2}$$

We say that $g \to F_g$ is a *deterministic process* if $\mathscr{H} = \mathscr{H}_1$. If $\mathscr{H} = \mathscr{H}_3$ the process is called *innovation process* and if $\mathscr{H} = \mathscr{H}_2$ the process is called *evanescent*.

The prediction problem is, then, nothing else than the characterization of the deterministic, evanescent and innovation parts of a process.

Let us define on G the real function ρ, by

$$\rho(g) = (F_g, F_1) \qquad (g \in G).$$

ρ is a positive definite function on G. Indeed, according to (9.5.1) we have

$$\Sigma\, c_j \bar{c}_i \rho(g_i^{-1} g_j) = \Sigma\, c_j \bar{c}_i (F_{g_i^{-1} g_j}, F_1) =$$

$$= \Sigma\, c_j \bar{c}_i (F_{g_j}, F_{g_i}) = \| \Sigma\, c_j F_{g_j} \|^2 \geqslant 0$$

for any finite system $c_1,\dots, c_n$ of complex numbers and any finite system $g_1,\dots, g_n$ of elements in G.

Following the Herglotz-Bochner-Weil theorem, there exists a positive measure μ on the dual X of the discrete group G, such that

$$\rho(g) = \int g \mathrm{d}\mu \qquad (g \in G),$$

where on the right handside g is viewed as character on X.

We now define the operator U on $\mathscr{K}$, with values in $L^2(\mathrm{d}\mu)$, by

$$UF_g = \bar{g} \qquad (g \in G). \tag{9.5.3}$$

Since

$$(UF_{g_1}, UF_{g_2}) = (\bar{g}_1, \bar{g}_2) = \int g_2\bar{g}_1 \,\mathrm{d}\mu = \rho(g_2 g_1{}^{-1}) =$$
$$= (F_{g_2 g_1^{-1}}, F_1) = (F_{g_2}, F_{g_1}),$$

U may be defined by (9.5.3) and is a unitary operator from $\mathscr{K}$ to $L^2(\mathrm{d}\mu)$

U realises an equivalence, from the point of view of the prediction theory, between the G-process $g \to F_g$ on $\mathscr{K}$ and the G-process $g \to F'_g = \bar{g}$, on $L^2(\mathrm{d}\mu)$. Indeed, if we write

$$M_g = \mathrm{clm}\,[\{F'_\gamma\colon\ \gamma \in G, g^{-1}\gamma \notin S\}]$$

then

$$M_g = U\mathscr{M}_g.$$

Let us remark that

$$M_g = \mathrm{clm}\,[\{\gamma\colon \gamma \in G, \gamma = gs, s \in S, s \neq 1\}].$$

Let A be the algebra generated by S in $C(X)$ and $f \to T_f$ be the natural representation of A on $H^2(\mathrm{d}\mu)$. We denote by m the Haar measure of X, viewed as the dual group of the group G.

Theorem 9.26. *The process $g \to F_g$ is deterministic (evanescent, innovation) if and only if the natural representation $f \to T_f$ of A on $H^2(\mathrm{d}\mu)$ is m-spectral (m-singular, m-completely non-spectral).*

Proof. Let $K = L^2(d\mu)$ and $H = H^2(d\mu)$. If we denote by U_g the operator of multiplication by g in $L^2(d\mu)$, then $g \to U_g$ is the unitary dilation of the semigroup $\{T_s\}_{s\in S}$ which corresponds to the natural representation $f \to T_f$ of A on $H^2(d\mu)$.

Let

$$K_1 = \bigcap_{g\in G} M_g$$

and

$$H = H_1 \oplus H_2 \oplus H_3$$

be the Wold decomposition of the semigroup of isometries $\{T_s\}_{s\in S}$.

Since

$$M_1 = \text{clm}\,[\{s: s \in S,\ s \neq 1\}]$$

there results $M_1 \subset H$, hence $K_1 \subset H$. By the definition of M_g we get that, for any $s \in S$, $s \neq 1$ and $h \in H$, we have $gsh \in M_g$. Therefore, $U_g K_1 \subset K_1$ for any $g \in G$ and, from (9.1.8) we obtain $K_1 \subseteq H_1$.

Conversely, since H_1 reduces any operator U_g to a unitary operator, for any $h \in H_1$ and $g \in G$ there exists $h_1 \in H_1$ such that $h = gsh_1$, with $\notin sS$, $s \neq 1$, i.e. $h \in M_g$. Then $H_1 \subset K_1$, hence $K_1 = H_1$.

Since the natural representation $f \to T_f$ is spectral if and only if $H_1 = H^2(d\mu) = L^2(d\mu)$, then $f \to T_f$ is spectral if and only if $K_1 = L^2(d\mu)$, that is if and only if the process $g \to F_g$ is deterministic.

Let now e_g be the projection of g on the orthogonal complement of M_g in K. As U_g is a unitary operator on K, there results

$$e_g = g \cdot e_1. \tag{9.5.4}$$

Assume now that $f \to T_f$ is m-singular. Then

$$1 \in \text{clm}\,[\bigcup_{s\neq 1} T_s H^2(d\mu)] = M_1.$$

Hence $e_1 = 0$ and, from (9.5.4) we get $e_g = 0$ for any $g \in G$. Therefore $K_3 = 0$. At the same time from $H_1 = 0$ there results $K_1 = 0$, that is $g \to F_g$ is an evancescent process.

Conversely, if the process is evanescent, then $K_1 = K_3 = \{0\}$. From $K_3 = 0$ there results $e_1 = 0$ and therefore

$$1 \in M_1 = \text{clm}\,[\bigcup_{s\neq 1} T_s H].$$

On the other hand, as $K_1 = 0$ implies $H_1 = 0$, we obtain that $f \to T_f$ is m-singular.

Let M be a subspace of H, doubly invariant to $f \to T_f$, such that

$$\text{clm}\,[\bigcup_{s \neq 1} T_s M] = M.$$

According to a previous remark, $M \subset M_1$. Then e_1 is orthogonal to M. Since $1 \in H^2(d\mu)$ and $M_1 \subset H^2(d\mu)$, there results $e_1 \in H^2(d\mu)$. We have

$$(m, e_s) = (m, se_1) = (\bar{s}m, e_1) = (P\bar{s}\, m, e_1) =$$

$$= (T_s^* m, e_1) = 0$$

$$(m, e_{\bar{s}}) = (m, \bar{s}e_1) = (sm, e_1) = (T_s m, e_1) = 0$$

for any $s \in S$, where P is the projection of $L^2(d\mu)$ on $H^2(d\mu)$. Therefore M is orthogonal to K_3.

If the process is innovation, then $M = \{0\}$, that is $f \to T_f$ is completely non-spectral.

If $f \to T_f$ is completely non-spectral then

$$H^2(d\mu) = \bigoplus_{s \in S} T_s N$$

where $N = M_1^\perp$ is one-dimensional.

Then, clearly,

$$L^2(d\mu) = \bigoplus_{g \in G} U_g N.$$

Let $k \in L^2(d\mu)$ be orthogonal to e_g for any $g \in G$. As $e_1 \in N$, $e_1 \neq 0$ and N is one-dimensional, we have $n_g = c_g e_1$ for any $n_g \in N$, with c_g constant, and

$$(k, gn_g) = (k, c_g g e_1) = c_g(k, ge_1) = c_g(k, e_g) = 0.$$

Hence k is orthogonal to $U_g N$ for any g and, from (9.5.4), we obtain $k = 0$. Thus, we have $L^2(d\mu) = K_3$, i.e. $g \to F_g$ is an innovation process.

The theorem is proved. ◇

In the case $G = Z$ is the additive group of integers and $S = Z^+$ the semigroup of positive integers, Theorem 9.24 gives exactly the classical prediction theorem for stationary discrete processes. The following two theorems are precising this fact. In this case we know that X is the unit circle $\{|z| = 1\}$ of the complex plane, m is the normalized Lebesgue measure on X and A the standard algebra on X.

Theorem 9.27. *Let A be the standard algebra on $X = \{|z| = 1\}$ and m the normalized Lebesgue measure on X. A positive measure μ on X is a Szegö measure if and only if it is a Szegö-total measure. This happens if and only if $\mathrm{d}\mu = h\mathrm{d}m$ with $h \in L^1(\mathrm{d}m)$ and*

$$\int \log h \, \mathrm{d}m > -\infty. \tag{9.5.5}$$

Proof. First observe that, in this case, the function 1 belongs to the $L^2(\mathrm{d}\mu)$-closure of A_μ if and only if $L^2(\mathrm{d}\mu) = H^2(\mathrm{d}\mu)$.

Let $\mathrm{d}\mu = h\mathrm{d}m + \mathrm{d}\mu_s$ be the Lebesgue decomposition of μ with respect to m. If E is the support of μ_s then, from the Szegö theorem and the above remark, there results

$$\chi_E L^2(\mathrm{d}\mu) = L^2(\mathrm{d}\mu_s) \subset H^2(\mathrm{d}\mu_s) \subset H^2(\mathrm{d}\mu). \tag{9.5.6}$$

Suppose μ is a Szegö measure. Then (9.5.6) yields $\mu(E) = 0$, that is $\mu_s = 0$. Hence $\mathrm{d}\mu = h\mathrm{d}m$. As $L^2(\mathrm{d}\mu) \neq H^2(\mathrm{d}\mu)$, once more from the Szegö theorem and the above remark, we get

$$\int \log h\mathrm{d}m > -\infty.$$

Now let $\mathrm{d}\mu = h\mathrm{d}m$ where h satisfies (9.5.5). The Szegö theorem yields $H^2(\mathrm{d}\mu) \neq L^2(\mathrm{d}\mu)$. Let E be a Borel set and

$$h_1 \in \chi_E L^2(\mathrm{d}\mu) \cap H^2(\mathrm{d}\mu).$$

Then $h^{-1}h_1\chi_E \in H^2(\mathrm{d}m)$ and we get $m(E) = 0$ or $m(X - E) = 0$, hence $\mu(E) = 0$ or $\mu(X - E) = 0$, i.e. μ is a Szegö-total measure. Since a total Szegö measure is a Szegö measure, the theorem is completely proved. ◇

Theorem 9.28. *Let $f \to T_f$ be the natural representation of the standard algebra A on $H^2(\mathrm{d}\mu)$. $f \to T_f$ is an m-completely non-spectral representation if and only if μ is a Szegö measure.*

Proof. There remains to prove that if μ is a Szegö measure, then

$$N = [\bigcup_{f \in A_m} T_f H^2(d\mu)]^{\perp} \neq 0.$$

Indeed, if $N = 0$ then 1 belongs to the $L^2(d\mu)$-closure of A_m and therefore the Szegö theorem yields $H^2(d\mu) = L^2(d\mu)$; but this contradicts the fact that μ is a Szegö measure.

The theorem is proved. ◇

Summing up all these results we now obtain the prediction theorem for discrete stationary processes.

Theorem 9.29. *Let $n \to F_n$ be a discrete stationary process given by a positive measure μ on the unit circle $X = \{|z| = 1\}$ of the complex plane,* and m the *Lebesgue measure on X. Let $d\mu = h dm + d\mu_s$, $h \in L^1(d\mu)$ be the Lebesgue decomposition of μ with respect to m. $n \to F_n$ is a deterministic process if and only if*

$$\int \log h \, dm = -\infty.$$

$n \to F_n$ is an innovation process if and only if $d\mu = h dm$ and

$$\int \log h \, dm > -\infty.$$

In the last case the prediction is made with a non-zero error e_1 given by Szegö formula

$$e_1 = exp\left[\frac{1}{2} \int \log h dm\right].$$

Notes

The first proof to Theorem 9.1 has been given by M. A. NAIMARK [1]. Other proofs as well as essential generalizations to non-commutative case and to semigroups are due to B. SZ.-NAGY [1], [2]. Theorem 9.5 has first been proved in the case of the additive group of integers and of the semigroups of positive integers by B. SZ.-NAGY [1], [2], by means of complex functions methods. The proof given here has been first done by I. HALPERIN, also in the case of integers group. W. MLAK [3] has rewritten the same proof for the case when G is totally ordered by S. The general form appears in I. SUCIU [7].

The theorems of canonical decomposition for semigroups of contractions have their origin in the works of B. Sz.-Nagy and C. Foiaş [1], [2], [6], where they are proved for the semigroup of positive integers or strongly continuous semigroups of real parameters. Important contributions in this direction have appeared in W. Mlak [1], [2], [3].

The characterization of the unitary space given by Theorem 9.9, in the general case, is given in I. Suciu [4].

Dilation theorem for semigroups of isometries has been given by T. Itô[1]. Wold type decomposition theorems for semigroups of isometries have their origin in H. Wold [1]. A first Wold type decomposition in three terms has been proved by H. Helson and D. Lowdenslager [2]. Theorems 9.11, 9.12 as well as the characterization of the semigroups of unilateral translations given by Theorems 9.13 and 9.14 are expounded in I. Suciu [9]. The results of paragraph 9.4 appear in I. Suciu [10] together with Theorem 9.24. The formulation of the prediction problem of paragraph 9.5 follows H. Helson and D. Lowdenslager [1]. Theorem 9.26 which establishes the connection between Helson and Lowdenslager prediction theorem and Theorem 9.24 is unpublished. Theorem 9.7 appears in C. Foiaş and I. Suciu [1].

Theorem 9.29 is the classical prediction theorem for discrete stationary processes (cf. J. L. Doob [1]).

Concerning prediction theorems see also G. Lumer [2].

CHAPTER 10

Some examples in the spectral theory of non-normal operators

10.1. The case of a single contraction

Let $X = \{z \in C \colon |z| = 1\}$ be the unit circle in the complex plane and A the standard algebra on X, that is the algebra of continuous functions on X which can be analytically extended in the interior of the unit disk $D = \{z \colon |z| < 1\}$. It is known that A is the uniform closure $P(X)$ of the polynomials in z and that A is a Dirichlet algebra on X.

At the same time, $S = \{1, z, z^2, \ldots\}$, where z is the function $f \in A$ defined by $f(x) = x$, $x \in X$, is a closed multiplicative system of inner functions in A, analytic free, which generates A. If we put $G = S\bar{S}$, then X is the dual of the discrete group G, the normalized Lebesgue measure is the Haar measure m of X, and

$$A = \{f \in C(X) \colon \int z^n f \mathrm{d}m = 0, \quad n = 1, 2, \ldots\}.$$

Let H be a Hilbert space and T a contraction on H, i.e. $T \in L(H)$, $\|T\| \leqslant 1$. We write $T_s = T^n$ for $s \in S$, $s = z^n$. We thus obtain the semigroup of contractions $\{T_s\}_{s \in S}$ on H.

Since $G = S \cup \bar{S}$ we obtain the map $g \to T_g$ of G in $L(H)$ by $T_{\bar{s}} = T_s^*$. According to Proposition 9.5, $g \to T_g$ is a positive definite extension of the semigroup $\{T_s\}_{s \in S}$ to G. Then from Corollary 9.3 there results that for any linear combination $\Sigma c_i s_i$ of elements in S, we have

$$\| \sum c_i T_{s_i} \| \leqslant \| \sum c_i s_i \|$$

that is for any polynomial $p(z)$

$$\|p(T)\| \leqslant \sup_{|z|=1} |p(z)| \tag{10.1.1}$$

which is von Neumann formula for a contraction.

Since $A = P(X)$, $f(T)$ can be constructed for any $f \in A$ by using (10.1.1). If we put $T_f = f(T)$ then it is easy to see that $f \to T_f$ is a representation of A on H such that $T_z = T$.

It is clear that any representation $f \to T_f$ of A on H may be obtained in the same way starting from the contraction $T = T_z$ of $L(H)$.

T is called completely *non-unitary* if the only subspace which reduces T to a unitary operator is $\{0\}$.

One easily verifies that T is unitary (completely non-unitary) if and only if $\{T_s\}_{s \in S}$ is a unitary (completely non-unitary) semigroup, hence if and only if $f \to T_f$ is a spectral (completely non-spectral) representation.

From Theorem 8.6 or 9.8 and Theorem 9.9 there results

Theorem 10.1. *Let T be a contraction on H. The space H has a unique decomposition of the form*

$$H = H_u \oplus H_c$$

where H_u reduces T to a unitary operator and H_c reduces T to a completely non-unitary contraction.

We have

$$H_u = \{h \in H\colon \|T^n h\| = \|T^{*n} h\| = \|h\|,\ n = 1, 2, \ldots\}.$$

Folowing Proposition 9.2 there exists a uniquely determined semispectral measure $(F(\sigma))_{\sigma \in B(X)}$ such that

$$(T^n h, k) = \int z^n d(F(x)h, k) \qquad (n = 0, 1, \ldots;\ h, k \in H).$$

Let $(\mu_{h,k})_{h,k\in H}$ be the semispectral family attached to the semispectral measure $(F(\sigma))_{\sigma\in B(X)}$ and $\mu_h = \mu_{h,h}$, $h \in H$.

From Theorem 8.16 and Theorem 9.7 there results

Theorem 10.2. *Let T be a completely non-unitary contraction on H and $(\mu_{h,k})_{h,k\in H}$ the semispectral family on X attached to T. For any $h \in H$ the measure μ_h is absolutely continuous with respect to the Lebesgue measure m on X and*

$$\int \log\left(\frac{d\mu_h}{dm}\right) dm > -\infty.$$

According to Corollary 9.5 the semigroup $\{T_s\}_{s\in S}$ generated by T admits a unitary dilation $g \to U_g$. The unitary operator $U = U_z$ is called the *unitary* dilation of T. Therefore

$$T^n h = PU^n h \qquad (h \in H,\ n = 1, 2, \ldots)$$

where P is the orthogonal projection of K (the dilation space) on H.

It is immediate that T is an isometry if and only if $\{T_s\}_{s\in S}$ is an isometric semigroup or if and only if $f \to T_f$ is a subspectral representation. In this case we have:

$$\operatorname{clm}[\bigcup_{s\sigma^{-1}\notin S^{-1}} T_\sigma^* T_s H] = \operatorname{clm}[\bigcup_{f\in A_m} T_f H] = TH.$$

Therefore $\{T_s\}_{s\in S}$ $(f \to T_f)$ is a quasy-unitary semigroup (m-spectral representation) if and only if it is unitary (spectral). At the same time $\{T_s\}_{s\in S}$ $(f\to T_f)$ is totally-nonunitary (m-completely non-spectral) if and only if it is completely nonunitary (completely non-spectral) and the singular semigroups (m-singular representations) do not appear.

The decomposition theorems of Wold Type (Theorem 8.25, Theorem 9.12) reduce in this case to the classical Wold theorem.

Theorem 10.3. (Wold). *Let T be an isometry on H. Then H admits a unique decomposition under the form*

$$H = H_u \oplus H_t$$

where H_u reduces T to a unitary operator and the restriction of T to H_t is unitarily equivalent to the shift (translation) operator in $H^2(N)$ with

$$N = [TH]^{\perp}.$$

Then Theorem 8.21 and Theorem 9.19 give

Theorem 10.4. *The isometry T on H is unitarily equivalent to the shift (translation) operator on $H^2(N)$, with N one-dimensional, if and only if T is not unitary and has no proper doubly invariant subspaces.*

10.2. Operators having spectral sets with connected complement

Let X be a compact set of the complex plane contained in the boundary of the unbounded connected component of its complement. Let $A = P(X)$ be the function algebra on X which is the uniform closure in $C(X)$ of the set of all polynomials in z (z is the function $f \in \in C(X)$ defined by $f(x) = x$ for $x \in X$). We recall that A is a Dirichlet algebra on X (Theorem 6.26). We also know that the maximal ideal space of X may be identified too with the polynomial convex hull $\hat{X}$ of X. We have

$$\hat{X} = X \cup G_1 \cup G_2 \cup \ldots$$

where $G_1, G_2, \ldots,$ are the bounded connected components of the complement of X. One easily verifies that $G_1, G_2, \ldots$ are the non point Gleason parts of $\hat{X}$ (relative to A) and any $x \in X$ forms a point Gleason part.

Theorem 10.5. *Let $f \to T_f$ be a representation of A on the Hilbert space H. Then H admits a decomposition under the form*

$$H = \bigoplus_{k=1}^{\infty} H_k \oplus H_s$$

where H_s, H_k, $h = 1, 2, \ldots$ are doubly invariant subspaces, $f \to T_f|H_s$ is spectral and for any $k = 1, 2, \ldots, f \to T_f|H_k$ is G_k-continuous.

Proof. Let $(G_\alpha)_{\alpha\in\mathscr{I}}$ be the family of all Gleason parts of $\hat{X}$ relative to A, and

$$H = \bigoplus_{\alpha\in\mathscr{I}} H_\alpha \oplus H_0$$

be the decomposition of H relative to $f \to T_f$ as follows from Theorem 8.28. Therefore H_0, H_α, $\alpha \in \mathscr{I}$ are doubly invariant subspaces, $f \to T_f|H_\alpha$ is G_α-continuous, $\alpha \in \mathscr{I}$, and $f \to T_f|H_0$ is singular. If $G_\alpha = \{x\}$ we denote H_α by H_x and write

$$H_s = \bigoplus_{x\in X} H_x \oplus H_0.$$

We now show that $f \to T_f|H_s$ is spectral. It is sufficient, for this, to prove that $f \to T_f|H_0$ and $f \to T_f|H_x$, $x \in X$, are spectral.

Let $(\mu_{h,k})_{h,k\in H}$ be the semi-spectral family attached to $f \to T_f$. If $h, k \in H_x$, then $\mu_{h,k}$ is absolutely continuous with respect to ε_x and, since $\mu_{h,k}(1) = (h, k)$, there results

(10.2.1) $$\mu_{h,k}(\varphi) = \varphi(x)\,(h, k) \qquad (h, k \in H_x).$$

Hence it is immediate that $(\mu_{h,k})_{h,k\in H_x}$ generates a representation of $C(X)$ on H_x which is an extension of $f \to T_f|H_x$, that is $f \to T_f|H_x$ is spectral.

Now let $h, k \in H_0$. We then know that $\mu_{h,k}$ is singular with respect to any representing measure for A. For any $f, g \in A$ we have

$$\int fg\, d\mu_{h,k} - \int f d\mu_{Tgh,k} = (T_{fg}h, k) - (T_f T_g h, k) = 0.$$

Therefore the measure $g d\mu_{h,k} - d\mu_{T_g h,k}$ is orthogonal to A. Since, obviously, it is singular with respect to any representing measure for A, from Proposition 6.34 we get

(10.2.2) $$\int fg d\mu_{h,k} = \int f d\mu_{T_g h,k} \qquad (f \in C(X),\ g \in A).$$

Let now $g \to T_g$ be the positive definite map of $C(X)$ in $L(H_0)$ given by the semispectral family $(\mu_{h,k})_{h,k\in H_0}$. Clearly $g \to T_g$ is an extension of $f \to T_f$. From (10.2.2) there follows

$$\int f \mathrm{d}\mu_{h,T_g^*k} - \int fg \, \mathrm{d}\mu_{h,k} = \int \mathrm{d}\mu_{T_fh,T_g^*k} - \int g \mathrm{d}\mu_{T_fh,k} =$$

$$= (T_fh, T_g^*k) - (T_gT_fh, k) = 0$$

for any $f \in A$ and $g \in C(X)$.

Therefore $\mathrm{d}\mu_{h,T_g^*k} - g \, \mathrm{d}\mu_{h,k}$ is an orthogonal measure to A and, as it is singular with respect to any representing measure of A, it is equal to zero.

Hence

$$\int fg \, \mathrm{d}\mu_{h,k} = \int f \mathrm{d}\mu_{h,T_g^*k} \qquad (h, k \in H_0, f, g \in C(X)). \tag{10.2.3}$$

From (10.2.3) there results

$$(T_{fg}h, k) = \int fg \, \mathrm{d}\mu_{h,k} = \int f \mathrm{d}\mu_{h,T_g^*k} =$$

$$= (T_fh, T_g^*k) = (T_gT_fh, k)$$

for any h, $k \in H_0$, i.e. $g \to T_g$ is multiplicative. Then $f \to T_f|H_0$ is a spectral representation.

Since a direct sum of spectral representations is a spectral representation, $f \to T_f|H_s$ is also spectral.

The theorem is proved. ◇

Let T be a bounded operator on H. Assume that for any polynomial p in z we have

$$\|p(T)\| \leqslant \sup_{x \in X} |p(x)|. \tag{10.2.4}$$

Then it is clear that for any $f \in A$ we can construct $f(T)$ and, writing $T_f = f(T)$, $f \in A$, we obtain a representation $f \to T_f$ of A on H, with $T_z = T$. One easily verifies that $f \to T_f$ is a spectral representation if and only it T is a normal operator and its spectrum is contained in X.

In this case the spectral measure attached to $f \to T_f$ is just the spectral measure of T.

Corollary 10.6. *Let T be a bounded linear operator on H which verifies (10.2.4). For any $k = 1, 2, \ldots$, let m_k be a representing measure with support in X, of a point in G_k.*

The space H admits a decomposition under the form

$$H = \bigoplus_{k=0}^{\infty} H_k$$

where H_0 reduces T to a normal operator T_0 with spectrum contained in X and its spectral measure singular with respect to m_k, $k = 1, 2, \ldots$, H_k, $k = 1, 2, \ldots$, reduces T to an operator which admits an absolutely continuous semispectral measure with respect to m_k.

10.3. Finite system of commuting contractions

Let $X = \prod_{1}^{n} \{z \in C: |z| = 1\}$ be the n-dimensional torus in C^n. We denote by $x = (z_1, z_2, \ldots, z_n)$ the points of X and by z_i the coordinate functions on X, i.e. $z_i(x) = z_i$, $i = 1, 2, \ldots, n$. Let S be the closed multiplicative system of inner functions on X generated by $z_1, z_2, \ldots, z_n$ and $G = S\bar{S}$. G is an abelian group, S is a sub-semigroup of G with $S \cap S^{-1} = \{1\}$ and X is the dual of the discrete group G. Let $A = A(S)$ be the function algebra on X generated by S. Therefore A is the closure in $C(X)$ of the algebra of polynomials in the variables $z_1, z_2, \ldots, z_n$.

Let $T_1, T_2, \ldots, T_n$ be a system of n commuting contractions on a Hilbert space H. If for any $s \in S$, $s = z_1^{k_1} z_2^{k_2} \ldots z_n^{k_n}$, we put $T_s = T_1^{k_1} T_2^{k_2} \ldots T_n^{k_n}$ we obtain a semigroup of contractions $\{T_s\}_{s \in S}$ on H.

Relative to the semigroup $\{T_s\}_{s \in S}$ it is natural to ask the following questions:

1) Does a unitary dilation $g \to U_g$ of $\{T_s\}_{s \in S}$ exist?

2) Does a representation $f \to T_f$ of A on H, which extends the semigroup $\{T_s\}_{s \in S}$ to A, exist?

It is clear that (2) is equivalent to the following inequality (von Neumann inequality)

(2′) for any polynomial $p(z_1, .. , z_n)$ in A we have

$$\|p(T_1, T_2, \ldots, T_n)\| \leqslant \sup_{x \in X} |p(x)|. \tag{10.3.1}$$

From Corollary 9.3 we get (1) ⇒ (2′).

In the following we prove that (1) is true for $n = 2$ (Ando Theorem). For $n = 3$ we give an example (Parrott's example) when (2′) is true and (1) is false.

We obtain, as a consequence, an example of representation of the algebra A which has no spectral dilation. *The assertion (2′) for $n \geqslant 3$ is an open question**).

We start with a lifting commutant theorem which is the version we need of a theorem of B. Sz.-Nagy—C. Foiaş.

Let T be a contraction on H and U acting on K be its minimal unitary dilation. Therefore H is a subspace of K, U a unitary operator on K and we have

$$T^n h = PU^n h \qquad (h \in H,\ n = 1, 2, \ldots)$$

$$K = \mathrm{clm}\,[\bigcup_{n=-\infty}^{\infty} U^n H]$$

where P is the orthogonal projection of K on H. Let $K_+ = \mathrm{clm}\,[\bigcup_{n=0}^{\infty} U^n H]$, and P_+ be the projection of K_+ on H. K_+ is obviously invariant to U and $V = U|K_+$ is an isometric operator on K_+. We also have

$$T^n h = P_+ V^n h \qquad (h \in H,\ n = 1, 2, \ldots). \tag{10.3.2}$$

This operator V will be called the isometrical dilation of T. Let

$$L = \overline{(V - T)\,H}.$$

*) For $n \geqslant 3$ and $\dim H \geqslant 5$ there are counterexamples for (2′) (cf. VARAPOULUS [1]).

From (10.3.2) there results

$$((V - T)h, h') = (Vh, h') - (Th, h') = (Th, h') - (Th, h') = 0$$

for any $h, h' \in H$, hence $H \perp L$. At the same time

$$(V^n(V - T)\,h, (V - T)\,h') = (V^{n+1}h, V\,h') -$$

$$- (V^nTh, Vh') - (V^{n+1}h, Th') + (V^nTh, Th') =$$

$$= (V^nh, h') - (V^{n-1}\,Th, h') - (V^{n+1}h, Th') + (V^nTh, Th') =$$

$$= (T^{n+1}h, h') - (T^{n+1}h, h') - (T^{n+1}h, Th') + (T^{n+1}h, Th') = 0$$

for any $n > 0$ and $h, h' \in H$. We used here (10.3.2) and the fact that V is an isometry on K_+. Therefore $V^nL \perp L$ for any $n > 0$ and, since V is an isometry on K_+, there results $V^{n_1}\,L \perp V^{n_2}\,L$ for any $n_1 \neq n_2$.

From the obvious relation

$$V^nh = T^nh + (V - T)\,T^{n-1}h + V(V - T)\,T^{n-2}h +$$

$$+ \ldots + V^{n-1}(V - T)\,h$$

there follows

(10.3.3) $$K_+ = H \oplus L \oplus VL \oplus V^2\,L \oplus \ldots$$

and from (10.3.3) one has

(10.3.4) $$P_+V = P_+VP_+.$$

Theorem 10.7. *Let Z be a bounded linear operator on H with $ZT = TZ$. There exists a bounded linear operator Y on K_+ with $YV = VY$ and such that*

(i) $T^n Z^m h = P_+ V^n Y^m h$ $(h \in H,\ n, m = 0, 1, 2..) \rightarrow$

(ii) $Y(K \ominus H) \subset H_+ \ominus H$

(iii) $\|Y\| = \|Z\|$.

Proof. We may assume $\|Z\| = 1$ and look for an Y of the form;

(10.3.5) $$YK = ZP_+k + B_0k + VB_1k + V^2B_2k + \dots \quad (k \in K_+)$$

where $B_0, B_1, B_2, \dots$ are bounded linear operators from K_+ into L. We shall determine $B_0, B_1, B_2, \dots$ such that Y satisfies (i), (ii), and (iii).

Let us define the operator B acting on K_+ with values in L, by

$$Bk = (V - T)\, ZP_+k \qquad (k \in K_+).$$

We have

$$\|Bk\|^2 = \|(V - T)\, ZP_+k\|^2 = ((V - T)\, ZP_+k,\ (V - T)\, ZP_+k) =$$

$$= ((V - T)\, ZP_+k,\ VZP_+k) = \|VZP_+k\|^2 - (TZP_+k,\ VZP_+k) =$$

$$= \|ZP_+k\|^2 - \|TZP_+k\|^2 = \|ZP_+k\|^2 - \|ZP_+Vk\|^2$$

where we used (10.3.2) and (10.3.4). Hence

(10.3.6) $$\|Bk\|^2 = \|ZP_+k\|^2 - \|ZP_+Vk\|^2 \qquad (k \in K_+).$$

Since $\|Z\| = 1$, the operator $I_{K_+} - P_+Z^*ZP_+$ is positive on K_+. Let D_0 be the positive square root of this operator. Then

$$D_0^2 = I_{K_+} - P_+Z^*ZP_+$$

$$\|D_0k\|^2 = (D_0^2k, k) = \|k\|^2 - \|ZP_+k\|^2 \qquad (k \in K_+)$$

and by (10.3.6) we have

$$\|Bk\|^2 = \|ZP_+k\|^2 - \|ZP_+Vk\|^2 \leqslant \|k\|^2 - \|ZP_+Vk\|^2 = \|Vk\|^2 - $$
$$- \|ZP_+Vk\|^2 = \|D_0Vk\|^2$$

that is

$$\|Bk\|^2 \leqslant \|D_0Vk\|^2 \qquad (k \in K_+). \tag{10.3.7}$$

The inequality (10.3.7) allows us to define a contraction C_0 from K_+ into L such that

$$B = C_0D_0V$$

as follows: we write $C_0D_0Vk = Bk$ on the subspace D_0VK_+ of K_+, then extend by continuity to the closure of D_0VK_+ and, for k in the orthogonal of $\overline{D_0VK_+}$ in K_+ we write $C_0k = 0$.

Let

$$B_0 = C_0D_0.$$

B_0 is an operator defined on K_+ with values in L.
We have $B_0V = B$ and

$$\|B_0k\|^2 = \|C_0D_0k\|^2 \leqslant \|D_0k\|^2 = \|k\|^2 - \|ZP_+k\|^2$$

that is

$$\|ZP_+k\|^2 + \|B_0k\|^2 \leqslant \|k\|^2. \tag{10.3.8}$$

We now construct by induction the sequence $B_0, B_1, B_2, \ldots$ of operators on K_+ with values in L such that

$$B_nV = B_{n-1} \qquad (n = 1, 2, \ldots) \tag{10.3.9}$$

and

$$\|ZP_+k\|^2 + \sum_{0 \leqslant n < p} \|B_nk\|^2 \leqslant \|k\|^2 \qquad (k \in K_+, p = 1, 2, \ldots) \tag{10.3.10}$$

Starting with B_0 defined above, suppose we constructed B_n, $0 \leqslant \leqslant n < p$ such that (10.3.9) for $n < p$, and (10.3.10) are satisfied. We get from (10.3.10) that $I_{K_+} - P_+Z^*ZP_+ - \sum_{0 \leqslant n < p} B_n^* B_n$ is a positive operator on K_+. Let D_p be its positive square root. Then

$$D_p^2 = I_{K_+} - P_+Z^*ZP_+ - \sum_{0 \leqslant n < p} B_n^* B_n$$

$$\|D_p k\|^2 = (D_p^2, k, k) = \|k\|^2 - \|ZP_+k\|^2 - \sum_{0 \leqslant n < p} \|B_n k\|^2, \quad (k \in K_+).$$

We have

$$\|ZP_+Vk\|^2 + \sum_{0 \leqslant n < p} \|B_n Vk\|^2 = \|ZP_+Vk\|^2 + \|Bk\|^2 + \sum_{0 < n < p} \|B_{n-1}k\|^2 \leqslant$$

$$\leqslant \|ZP_+k\|^2 + \sum_{0 \leqslant n < p-1} \|B_n k\|^2 = \|ZP_+k\|^2 + \sum_{0 \leqslant n < p} \|B_n k\|^2 -$$

$$- \|B_{p-1}k\|^2 \leqslant \|k\|^2 - \|B_{p-1}k\|^2 = \|Vk\|^2 - \|B_{p-1}k\|^2$$

for any $k \in K_+$. We used here relation $B_0V = B$, (10.3.6), (10.3.9) for $n < p$ and (10.3.10). Therefore

$$\|B_{p-1}k\|^2 \leqslant \| k\|^2 - \|ZP_+Vk\|^2 - \sum_{0 \leqslant n < p} \|B_n Vk\|^2 = \|D_p Vk\|^2$$

thus

$$\|B_{p-1}k\|^2 \leqslant \|D_p Vk\|^2 \qquad (k \in K_+). \tag{10.3.11}$$

A similar argument to that used in constructing C_0, allows us to construct, by (10.3.11), a contraction C_p from K_+ into L, such that

$$B_{p-1} = C_p D_p V.$$

Let

$$B_p = C_p D_p.$$

Then

$$B_p V = B_{p-1}$$

and

$$\|B_p k\|^2 = \|C_p D_p k\|^2 \leqslant \|D_p k\|^2 = \|k\|^2 - \|ZP_+ k\|^2 - \sum_{0 \leqslant n < p} \|B_n k\|^2$$

for any $k \in K_+$, hence

$$\|ZP_+ k\|^2 + \sum_{0 \leqslant n < p+1} \|B_n k\|^2 \leqslant \|k\|^2 \qquad (k \in K_+).$$

Thus we get the operators $B_0, B_1, \ldots, B_p$ from K_+ into L such that relations (10.3.9) and (10.3.10) are satisfied for $n < p + 1$.

Therefore relation (10.3.5) defines an operator Y on K. We have

$$\|Yk\|^2 = \|ZP_+ k\|^2 + \sum_{n=0}^{\infty} \|B_n k\|^2 \leqslant \|k\|^2$$

hence $\|Y\| \leqslant 1$. According to the definition of Y and to (10.3.3) there results

$$Y(K_+ \ominus H) \subset K_+ \ominus H$$

$$ZP_+ = P_+ Y,$$

hence we have

$$Z^n P_+ = Z^{n-1} Z P_+ = Z^{n-1} P_+ Y = Z^{n-2} Z P_+ Y = Z^{n-2} P_+ Y^2 =$$

$$= \ldots = P_+ Y^n.$$

Thus

$$T^n Z^m h = P_+ V^n Z^m h = P_+ V^n P_+ Y^m h = P_+ V^n Y^m h$$

$$(h \in H;\ n, m = 0, 1, 2, \ldots).$$

At the same time

$$1 = \|Z\| = \|ZP_+\| = \|P_+ Y\| \leqslant \|Y\|$$

and, therefore,

$$\|Y\| = \|Z\| = 1.$$

We now calculate

$$VY - YV = V\left(ZP_+ + \sum_{n=0}^{\infty} V^n B_n\right) - \left(ZP_+ + \sum_{n=0}^{\infty} V^n B_n\right)V =$$

$$= VZP_+ + \sum_{n=0}^{\infty} V^{n+1} B_n - ZP_+ V - \sum_{n=0}^{\infty} V^n B_n V = VZP_+ - ZP_+ V -$$

$$- B_0 V + \sum_{n=0}^{\infty} V^n (B_{n-1} - B_n V) = VZP_+ - ZTP_+ - B_0 V =$$

$$= VZP_+ - TZP_+ - B_0 V = (V - T)\, ZP_+ - B_0 V = B - B_0 V = 0$$

so

$$YV = VY.$$

The theorem is proved. ◇

Proposition 10.8. *Let Z be a bounded linear operator on H with $ZT = TZ$, Y a bounded linear operator on K_+ with $YV = VY$ and suppose the assertion (i), (ii), (iii) of Theorem 10.7 are satisfied. If Z is an isometry on H then Y is an isometry on K_+ and $YH \subset H$, $YL \subset L$. If Z is unitary on H then Y is unitary on K_+.*

Proof. Suppose that Z is an isometry on H.

For any $h \in H$ we have

$$\|h\| \geqslant \|Yh\| \geqslant \|P_+ Yh\| = \|Zh\| = \|h\|.$$

Thus

$$\|Yh\| = \|P_+ Yh\|$$

hence

$$Yh = P_+ Yh = Zh.$$

Thus

$$YH \subset H; \; Z = Y|H.$$

Since $Y(K_+ \ominus H) \subset K_+ \ominus H$ there results that H reduces Y. Now

$$YV^n(V - T)h = V^n Y(V - T)h = V^n(VY - YT)h$$

and since

$$YTh = ZTh = TZh$$

we have

$$YV^n(V - T)h = V^n(V - T)Zh.$$

Thus

$$YV^n L \subset V^n L.$$

It remains to be proved that $Y|V^n L$ is isometric.
But

$$\|Y(V - T)h\|^2 = \|(V - T)Zh\|^2 = \|Zh\|^2 - \|ZTh\|^2 =$$

$$= \|h\|^2 - \|Th\|^2 = \|(V - T)h\|^2.$$

Hence

$$\|Yl\| = \|l\| \qquad (l \in L).$$

Now

$$\|YV^n l\| = \|V^n Yl\| = \|l\| = \|V^n l\|,$$

therefore $Y|V^n L$ is isometric. Consequently Y is an isometry on K_+.
If Z is unitary, then Z is an isometry and $ZH = H$. Therefore Y is an isometry on K_+ and $Yh = Zh$, $h \in H$. Let $k \in K_+$ be of the form

$$k = \sum_{n=0}^{p} V^n h_n \qquad (h_n \in H).$$

There exist $h'_n \in H$ such that $h_n = Zh'_n$, $0 \leqslant n \leqslant p$, so

$$k = \sum_{n=0}^{p} V^n Z h'_n = \sum_{n=0}^{p} V^n Y h'_n = Y \sum_{n=0}^{p} V^n h'_n.$$

Since the elements of this form are dense in K_+ and Y is an isometry on K_+, there follows $YK_+ = K_+$ that is Y is unitary on K_+.

The proposition is proved. ◇

Theorem 10.9. (lifting commutants theorem). *Let T be a contraction on H and U its unitary dilation acting on K. For any bounded linear operator Z on H which commutes with T there exists a bounded linear operator Y on K which commutes with U such that*

(i) $T^n Z^m h = PU^n Y^m h \qquad (h \in H, m, n = 0, 1, 2, \ldots)$

(ii) $YK_+ \subset K_+$, $Y(K_+ \ominus H) \subset K_+ \ominus H$

(iii) $\|Y\| = \|Z\|$.

If Z is unitary then Y is unitary and $YH \subset H$, $YL \subset L$.

Proof. Let Y_+ be the operator on K_+ obtained by the preceding theorem for operators T and Z on H. Since Y_+ commutes with V then Y_+^* commutes with V^*. If we apply once more the preceding theorem for V^*, Y_+^* and take into account that the isometrical dilation of V^* is U^* we obtain an operator Y^* on K which commutes with U^* and satisfies

$$V^{*n} Y_+^{*m} k = P_{K_+} U^{*n} Y^{*m} k \quad (k \in K, n, m = 1, 2, \ldots)$$

$$Y^*(K \ominus K_+) \subset K \ominus K_+$$

$$\|Y^*\| = \|Y_+^*\|$$

where P_{K_+} is the projection of K on K_+. It is easy to see that $Y = (Y^*)^*$ is the desired operator. ◇

Theorem 10.10. *For $n = 2$ the semigroup $\{T_s\}_{s \in S}$, constructed at the beginning of this paragraph, has a unitary dilation.*

Proof. Let T_1, T_2 be the two commuting contractions on H. Let U_1' acting on K_1 be the unitary dilation of T_1. According to Theorem 10.9 there exists a contraction T_2' on K_1 such that

$$T_2'U_1' = U_1'T_2$$

$$T_1^nT_2^mh = P_1U_1'^nT_2'^mh \qquad (h \in H, n, m = 1, 2, \ldots)$$

where P_1 is the projection of K_1 onto H.

Now let U_2 acting on K be the unitary dilation of T_2'. If we apply once more Theorem 10.9 we get the unitary operator U_1 on K which commutes with U_2 such that

$$U_1'^nT_2'^mk = P_2U_1^nU_2^mk \quad (k \in K_+, \ n = 0, 1, 2, \ldots)$$

where P_2 is the projection of K on K_1. Let P be the projection of K on H. We have

$$T_1^nT_2^mh = P_1U_1'^nT_2'^mh = P_1P_2U_1^nU_2^mh = PU_1^nU_2^mh$$

that is

$$T_1^nT_2^mh = PU_1^nU_2^mh \qquad (h \in H, n, m = 0, 1, 2, \ldots).$$

If we put $U_g = U_1^nU_2^m$ for any $g \in G$, $g = z_1^nz_2^m$, $n, m = 0, \pm 1, \pm 2, \ldots$, then $g \to U_g$ is clearly a unitary representation of G on K which dilates the semigroup $\{T_s\}_{s \in S}$. ◇

Corollary 10.11. *For $n = 2$ the inequality (10.3.1) holds, that is for any two commuting contractions T_1, T_2 on H and any polynomial in two variables $p(z_1, z_2)$ we have*

$$\|p(T_1, T_2)\| \leqslant \sup_{|z_1| = |z_2| = 1} |p(z_1, z_2)|.$$

We now give an example of three commuting contractions T_1, T_2, T_3 on H for which the semigroup $\{T_s\}_{s \in S}$ generated by them has no unitary dilation but verifies inequality (10.3.1). This example is mainly due to Parrott ([1]).

Theorem 10.12. *For $n = 3$ there are semigroups $\{T_s\}_{s \in S}$ which satisfy (10.3.1) but do not admit unitary dilations.*

Let H_0 be a Hilbert space on which there exist two non-commuting unitary operators A_1, A_2

$$A_1A_2 \neq A_2A_1.$$

Let us denote by A_3 the identity operator on H_0.

We write $H = H_0 \oplus H_0$ and consider on H operators T_i, $i = 1. 2. 3$ defined by the matrix

$$T_i = \begin{pmatrix} 0 & 0 \\ A_i & 0 \end{pmatrix}.$$

It is obvious that $\|T_i\| \leqslant 1$, $T_iT_j = T_jT_i = 0$ for $i, j = 1, 2, 3$. Hence T_1, T_2, T_3 is a system of three commuting contractions on H. Assume there exists a Hilbert space K and the commuting unitary operators U_1, U_2, U_3 on K such that $H \subset K$ and

$$\text{(10.3.12)} \qquad T_ih = PU_ih \qquad (h \in H,\ i = 1, 2, 3)$$

where P is the orthogonal projection of K on H.

For $h \in H$, $h = [h_0, 0]$, $h_0 \in H_0$, we have

$$\|T_ih\| = \|[0, A_ih_0]\| = \|A_ih_0\| = \|h_0\| = \|h\| = \|U_ih\|$$

and from (10.3.12) we get $T_ih = U_ih$.

Then

$$U_i[h_0, 0] = T_i[h_0, 0] = [0, A_ih_0]$$

$$U_j^{-1}U_i[h_0, 0] = U_j^{-1}[0, A_ih_0] = U_j^{-1}[0, A_jA_j^{-1}A_ih_0] =$$

$$= U_j^{-1}U_j[A_j^{-1}A_ih_0, 0] = [A_j^{-1}A_ih_0, 0]$$

$$U_kU_j^{-1}U_i[h_0, 0] = U_k[A_j^{-1}A_ih_0, 0] = [0, A_kA_j^{-1}A_ih_0].$$

Since U_1, U_2, U_3 are commuting operators, there results

$$[0, A_1A_2h_0] = U_1U_3^{-1}U_2[h_0, 0] = U_2U_3^{-1}U_1[h_0, 0] = [0, A_2A_1h_0]$$

that is $A_1A_2 = A_2A_1$ which is impossible. Therefore, the semigroup $\{T_s\}_{s\in S}$ has no unitary dilation.

Let us now show that T_1, T_2, T_3 verify inequality (10.3.1). Let

$$p(z_1, z_2, z_3) = a_0 + a_1z_1 + a_2z_2 + a_3z_3 + \dots$$

be a polynomial in variables z_1, z_2, z_3. Since $T_iT_j = 0$ for any $i, j = 1, 2, 3$, there follows

$$p(T_1, T_2, T_3) = a_0 + a_1T_1 + a_2T_2 + a_3T_3$$

that is

$$p(T_1, T_2, T_3) = \begin{pmatrix} a_0 & 0 \\ \sum_{i=1}^{3} a_iA_i & a_0 \end{pmatrix}.$$

Let H' be the subspace generated by $\{1, z\}$ in $L^2(d\mu)$ where μ is the Lebesgue measure on the unit circle $\{z \in C: |z| = 1\}$ of the complex plane, and A the operator on H' given by the matrix

$$A = \begin{pmatrix} |a_0| & 0 \\ \sum_{i=1}^{3} |a_i| & |a_0| \end{pmatrix}.$$

One easily verifies that $\|p(T_1, T_2, T_3)\| \leqslant \|A\|$. On the other hand if P' is the projection of L^2 on H' and q a one-variable polynomial

$$q(z) = |a_0| + \sum_{i=1}^{3} |a|z + \dots$$

then

$$Ah = P'qh \qquad (h \in H).$$

Hence

$$\|Ah\| = \|P'qh\| \leqslant \|qh\| = \int |q|^2|h|^2 d\mu \leqslant \sup_{|z|=1} |q(z)| \; \|h\|.$$

that is

$$\|A\| \leqslant \sup_{|z|=1} |q(z)|.$$

Since

$$p(z_1, z_2, z_3) = \frac{a_0}{|a_0|}\left(|a_0| + \sum_{i=1}^{3} |a_i| \frac{\bar{a}_0 a_i}{|a_0 a_i|} z_i + \ldots\right)$$

with the convention $\frac{a_i}{|a_i|} = 1$ if $a_i = 0$, there results

$$\sup_{|z_i|=1} |p(z_1, z_2, z_3)| = \sup_{|z_i|=1} \left| |a_0| + \sum_{i=1}^{3} |a_i| z_i + \ldots \right| \geqslant$$

$$\geqslant \sup_{\substack{z_1=z_2=z_3=z \\ |z|=1}} \left| |a_0| + \sum_{i=1}^{3} |a_i| z_i + \ldots \right| =$$

$$= \sup_{|z|=1} \left| |a_0| + \sum_{i=1}^{3} |a_i| z + \ldots \right| \geqslant \|A\| \geqslant \|p(T_1, T_2, T_3)\|$$

that is

$$\|p(T_1, T_2, T_3)\| \leqslant \sup_{|z_i|=1} |p(z_1, z_2, z_3)|.$$

Corollary 10.13. *There are representations of a function algebra which admit no spectral dilation.*

Notes

The results presented in paragraph 10.1 have been mainly taken from the contraction theory on Hilbert spaces derived by B. Sz.-Nagy and C. Foiaş in a series of common works and unitarily exposed in the book of B. Sz.-Nagy and C. Foiaş [6].

The second part of Theorem 10.2 has been proved by W. Mlak [2]. Theorem 10.3 is the classical theorem of H. Wold [1].

Theorems of paragraph 10.2 lead to the principal result obtained by D. Sarason [1].

Theorem 10.7 appears in a more general form in B. Sz.-Nagy and C. Foiaş [5] while Proposition 10.8 and Theorem 10.9 follow simply from it.

Theorem 10.10 has been proved by T. Ando [1] and the example contained in Theorem 10.2 has been essentially constructed by S. Parrott [1].

References

AHERN, P. R., SARASON, D.,

The H^p-spaces of a class of function algebras. *Acta Math.* 117 (1967), 123–163

ANDO, T.,

[1] On a pair of commutative contractions. *Acta Sci. Math.* 24 (1963), 88–90

ARENS, R., CALDERON, A.,

[1] Analytic functions of several Banach algebra elements. *Ann. Math.* 62 (1955), 204–216

ARENS, R., SINGER, I. M.,

[1] Function values as boundary integrals. *Proc. Amer. Math. Soc.* 5 (1954), 735–745

[2] Generalized analytic functions. *Trans. Amer. Math. Soc.* 81 (1956), 379–393

ARVESON, W. B.,

[1] On subalgebras of C*-algebras. *Bull. Amer. Math. Soc.* 75 (1969), 790–794

[2] Subalgebras of C*-algebras, *Acta Math.* 123 (1969), 141–224

BAUER, H.,

[1] Silovsher Rand und Dirichletsches Problem. *Ann. Inst. Fourier (Grenoble)*, 11 (1961), 89–133

BEAR, H. S.,

[1] A strong maximum modulus principle for maximal function algebras. *Trans. Amer. Math. Soc.* 92 (1959), 465–469

[2] A geometric characterization of Gleason parts. *Proc. Amer. Math. Soc.* 16 (1965), 407–412

BEURLING, A.,

[1] On two problems concerning linear transformations in Hilbert space. *Acta Math.* 81 (1949), 239–255

BISHOP, E.,

[1] A minimal boundary for function algebras. *Pacific J. Math.* 9 (1959), 629–642

[2] A generalization of the Stone-Weierstrass theorem. *Pacific J. Math.* 11 (1961), 777–783

[3] A general Rudin-Carleson theorem. *Proc. Amer. Math. Soc.* 13 (1962), 140-143

[4] Representing measures for points in a uniform algebra. *Bull. Amer. Math. Soc.* 70 (1964), 121–122

Bishop, E., de Leeuw, K.,

[1] The representations of linear functionals by measures on sets of extreme points. *Ann. Inst. Fourier.* 9 (1959), 305–331

Boboc, N., Cornea, A.,

[1] Convex cones of lower semicontinuous functions on compact spaces. *Rev. Roum. Math. pures et appl.* 12 (1967), 471–525

Brelot, M.,

[1] Lectures on potential theory. Tata Inst. Fund. Research, Bombay (1960)

Bourbaki, N.,

[1] *Intégration*, Chap. I–IV. Actual. sci et industr., 1175, Hermann, Paris (1952)

[2] *Espaces vectoriels topologiques.* Chap. I–II. Actual. sci. et industr., 1189, Hermann, Paris (1953)

Carleson, L.,

[1] Interpolations by bounded analytic functions and the corona problem. *Ann. Math.* 76 (1962), 547–559

[2] Mergelyan's theorem on uniform approximations. *Math. Scand.* 15 (1964), 167–175

Choquet, G.,

[1] Existence et unicité des représentations intégrales au moyen des points extrémaux dans les cônes convexes. *Seminaire Bourbaki*, no. 139 (1956). (Institut H. Poincaré, Paris)

Choquet, G., Meyer, P. A.,

[1] Existence et unicité des représentations intégrales dans les connexes compacts quelconques. *Ann. Inst. Fourier*, 13 (1963), 139–154

Dinculeanu, N.,

[1] *Vector measures.* Deutscher Verlag der Wissenschaften, Berlin, 1966

Doob, J. L.,

[1] *Stochastic processes.* Wiley, New York, 1953

Dunford, N., Schwartz, J.,

[1] *Linear operators.* Part I. Wiley, Interscience, New York, 1958

Edwards, D. A.,

[1] On the representation of certain functionals by measures on the Choquet boundary. *Ann. Inst. Fourier*, 13 (1963), 111–121

Foiaş, C.,

[1] Certaines applications des ensembles spectraux. I. Mesure harmonique-spectrale. *Studii şi cerc. mat.* 10 (1959), 365–401 (Romanian)

[2] La maximalité de l'espace H^∞ dans le calcul fonctionnel. *Analele Univ. Timişoara* (ser. mat.-fiz.), 2 (1964), 77–81 (Romanian)

[3] Modèles fonctionnels, liaison entre les théories de la fonction charactéristique et de la dilatation unitaire. *Deuxième Colloque sur l'analyse fonctionnelle* (Liège, 1964), 63–76

[4] Spectral and semispectral measures. *Studii şi cerc. mat.* 18 (1966), 7–56 (Romanian)

FOIAŞ, C., SUCIU, I.,

[1] Szegö-measures and spectral theory in Hilbert spaces. *Rev. Roum. Math. pures et appl.* 11 (1966), 147–159

[2] On the operator representations of logmodular algebras. *Bull. Acad. Polon. Sci.* 16 (1968), 505–509

GAMELIN, T.,

[1] Restriction of subspaces of *C(X)*, *Trans. Amer. Math. Soc.* 112 (1964), 278–286

[2] *Uniform algebras.* Prentice-Hall, Englewood Cliffs (1969)

GAŞPAR, D.,

[1] Some properties of H^p-spaces in the case of function algebras. *Analele Univ. Timişoara* (ser. mat-fiz.), 6 (1968), 149–157 (Romanian)

[2] On representations of function algebras. *Acta Sci. Math.* 31 (1970), 339–346

GELFAND, I., RAIKOV, D., SHILOV, G.,

[1] *Commutative normed rings.* Fizmatgiz, Moscow, 1960 (Russian)

GLEASON, A.,

[1] Function algebras. *Seminars on Analytic Functions,* vol. II. Princeton 1957, 213–216

GLICKSBERG, I.,

[1] Measures orthogonal to algebras and sets of antisymmetry. *Trans. Amer. Math. Soc.* 105 (1962), 415–435

[2] Function algebras with closed restrictions. *Proc. Amer. Math. Soc.* 14 (1963), 158–161

[3] The abstract F. and M. Riesz theorem. *J. Functional Analysis,* 1 (1967), 109–122

GLICKSBERG, I., WERMER, J.,

[1] Measures orthogonal to a Dirichlet algebra. *Duke Math. J.* 30 (1963), 661–666

GORIN, E. A.,

[1] Maximal subalgebras of commutative Banach algebras with involution. *Mat. Zamet.* 1 (1967), 175–178 (Russian)

[2] Commutative Banach algebras generated by the group of unitary elements. *Funk. Analiz. Priloj.* 1 (1967), 86–87 (Russian)

GUNNING, E., ROSSI, H.,

[1] *Analytic Functions of Several Complex Variables.* Prentice-Hall, Inc. Englewood Cliffs, New Jersey, 1965

Halmos, P. R.,

[1] *Introduction to Hilbert space and the theory of spectral multiplicity*. Chelsea Publishing Company, New York, 1951

[2] Normal dilations and extensions of operators. *Summa Brasil. Math.* 2 (1950), 125–134

[3] Shifts on Hilbert spaces. *J. reine angew. Math.* 208 (1961), 102–112

[4] *A Hilbert space problem book*. Van Nostrand, Princeton–Toronto–London, 1967

Halperin, I.,

[1] Sz.-Nagy-Brehmer dilations. *Acta Sci. Math.* 23 (1962), 279–289

Helson, H.,

[1] *Lectures on invariant subspaces*. Academic Press, New York–London, 1964

Helson, H., Lowdenslager, D.,

[1] Prediction theory and Fourier series in several variables. *Acta Math.* 99 (1958), 165–202

[2] Prediction theory and Fourier series in several variables. II. *Acta Math.* 106 (1961), 175–213

Helson, H., Quigley, F.,

[1] Maximal algebras of continuous functions. *Proc. Amer. Math. Soc.* 8 (1957), 111–114

Hoffman, K.,

[1] Analytic functions and logmodular Banach algebras. *Acta Math.* 108 (1962), 271–317

[2] *Banach Spaces of Analytic Functions*. Prentice-Hall, Inc., Englewood Cliffs, New Jersey, 1962

[3] Lectures on supnorm algebras. *Summer School on Topological Algebra Theory*, Bruges, 1966, 1–74.

[4] Bounded analytic functions and Gleason parts. *Ann. Math.* 86 (1967), 74–111

Hoffman, K., Rossi, H.,

[1] Functions theory and multiplicative linear functionals. *Trans. Amer. Math. Soc.* 116 (1965), 530–543

Hoffman, K., Singer, I. M.,

[1] Maximal subalgebras of continuous functions. *Acta Math.* 103 (1960), 217–241

Hoffman, K., Wermer, J.,

[1] A characterization of C(X). *Pacific J. Math.* 12 (1962), 941–944

Ionescu-Tulcea, C. T.,

[1] *Hilbert spaces*. Ed. Acad. R.P.R. Bucharest, 1954 (Romanian)

Itô, T.,

[1] On the commutative family of subnormal operators. *J. Fac. Sci. Hokkaido Univ.* (1), 14 (1958), 311–312

König, H.,

[1] Zur abstrakten Theorie der analytischen Functionen. *Math. Zeitschr.* 88 (1965), 536–543

[2] Zur abstrakten Theorie der analytischen Functionen. II. *Math. Ann.* 163 (1966), 9–17

[3] Zur abstrakten Theorie der analytischen Funktionen. III. *Arch. Math.* 18 (1967), 273–284

[4] Lectures on abstract H^p-theory. *Summer School on Topological Algebra Theory.* Bruges, 1966, 75–127

Lumer, G.,

[1] Analytic functions and the Dirichlet problem. *Bull. Amer. Math. Soc.* 70 (1964), 98–104

[2] *Algèbres de fonctions et espaces de Hardy.* Lecture Notes in Mathematics 75, Springer-Verlag, Berlin, Heidelberg, New York, 1968

Mergelyan, S.,

[1] Uniform approximation for functions of a complex variable, *Uspehi Mat. Nauk*, nr. 7 (1952), 31–122, *Amer. Math. Soc. Transl.* 101 (1954)

Merrill, S.,

[1] Maximality of certain algebras H^∞(dm). *Math. Zeitschr.* 106 (1968), 261–266

Mlak, W.,

[1] Representations of some algebras of generalized analytic functions. *Bull. Acad. Polon. Sci. (sér. math. astr., phys.)*, 13 (1965), 211–214

[2] Unitary dilations of contraction operators, *Rozprawy mat.* 46 (1965), 1–88

[3] Unitary dilations in case of ordered groups. *Ann. Pol. Math.* 17 (1965), 321–328

[4] Positive definite contraction valued functions. *Bull. Acad. Polon. Sci, (sér. math., astr., phys.)*, 15 (1967), 509–512

[5] A note on Szegö type properties of semi-spectral measures. *Studia Math.* 31 (1968), 241–251

[6] Decompositions and extensions of operator valued representations of function algebras. *Acta Sci. Math.* 30 (1969), 181–193

Mochizuki, N.,

[1] A characterization of the algebra of generalized analytic functions. *Tôh. Math. J.* 16 (1964), 313–319

[2] Isometry between H^p(dm) and the Hardy class H^p. *Tôh. Math. J.* 18 (1966) 311–315

[3] Correction to "Isometry between $H^p(dm)$ and the Hardy class H^p". *Tôh. Math. J.* 19 (1967), 373

MUHLY, P. S.,

[1] A structure theory for isometric representations of a class of semigroups, *J. Reine Angew. Math.* 255 (1972), 135–154

NAIMARK, M. A.,

[1] Positive definite operator functions on a commutative group. *Izvestia Akad. Nauk SSSR*, 7 (1943), 237–244 (Russian)

[2] On a representation of additive operator set functions. *Doklady Akad. Nauk SSSR*, 41 (1943), 359–361 (Russian)

[3] *Normed Rings*. GITTL., Moscow, 1956

NICOLESCU, M.,

[1] *Mathematical Analysis*, I, II, III. Ed. tehnică, Bucureşti, 1957, 1958, 1960 (Romanian)

PARROTT, S.,

[1] Unitary dilations for commuting contractions. *Pacific J. Math.* 34 (1970), 481–490

PHELPS, R.,

[1] *Lectures on Choquet's theorem*. Van Nostrand, Princeton, 1966

RIESZ, F., SZ.-NAGY, B.,

[1] *Leçons d'analyse fonctionnelle* (IVth ed.), Akadémiai Kiado, Budapest—Gauthier-Villars, Paris, 1965

ROSSI, H.,

[1] The local maximum modulus principle. *Ann. Math.* 72 (1960), 1–11

RUDIN, W.,

[1] *Fourier Analysis on Groups*. Interscience, New York, 1962

SARASON, D.,

[1] On spectral sets having connected complement. *Acta Sci. Math.* 26 (1965), 289–299

[2] Generalized interpolation in H^∞. *Trans. Amer. Math. Soc.* 127 (1967), 179–203

SHILOV, G. E.,

[1] Function rings with uniform convergence. *Ukr. Math. J.* 4 (1951), 404–411 (Russian)

[2] On the decomposition of a commutative normed ring into a direct sum of ideals. *Matem. Sb.* 32 (1965), 353–364 (Russian)

[3] Analytic functions on normed rings. *Usp. Mat. Nauk.* 15 (1960), 3 (Russian)

STINESPRING, W. F.,

[1] Positive functions on C*-algebras. *Proc. Amer. Math. Soc.* 6 (1955), 211–216

STOILOW, S.,

[1] *The theory of functions of a complex variable.* vol. I. Ed. Acad. R.P.R., Bucureşti, 1954 (Romanian)

STOILOW, S., (in collaboration with) ANDREIAN CAZACU, C.,

[1] *The theory of functions of a complex variable.* vol. II. Ed. Acad. R.P.R., Bucureşti, 1958 (Romanian)

SUCIU, I.,

[1] Natürliche Erweiterung der kommutativen Banach-algebra. *Rev. Math. pures et appl.* 7 (1962), 483—491

[2] Bruchalgebra der Banach-algebren. *Rev. Math. pures et appl.* 8 (1963), 313—316

[3] Analytic Banach algebras. *Studii şi cerc. mat.* 16 (1964), 55—76 (Romanian)

[4] Spectral dilatable representations of commutative Banach algebras. *Studii şi cerc. mat.* 16 (1964), 1211—1220 (Romanian)

[5] On the algebras generated by the inner functions. *Rev Roum. Math. pures et appl.* 10 (1965), 1423—1430

[6] Interpolation sequences and maximal ideal space of H^∞. *Studii şi cerc. mat.* 18 (1966), 1029—1038 (Romanian)

[7] Unitary dilation in case of partially ordered groups. *Bull. Acad. Polon. Sci. (sér. math. astr. phys.)*, 15 (1967), 271—275

[8] H^p-spaces in case of function algebras. *Studii şi cerc. mat.* 19 (1967), 757—801 (Romanian)

[9] On the semigroups of isometries. *Studia Math.* 30 (1968), 101—110

[10] Spectral theory for operator representations of function algebras. *Studi. şi cerc. mat.* 20 (1968) (Romanian)

[11] The Wold decomposition for the subspectral representations of function algebras. *Rev. Roum. Math. pures et appl.* 16 (1971), 281—296

SZEGÖ, G.,

[1] Über die Randtewer analytischer Funktionen. *Math. Ann.* 84 (1921), 232—244

SZ.-NAGY, B.,

[1] Transformation de l'espace de Hilbert, fonctions de type positif sur un groupe. *Acta Sci. Math.* 15 (1954), 104—114

[2] Prolongements des transformations de l'espace de Hilbert qui sortent de cet espace. Append. to F. RIESZ and B. SZ.-NAGY [1].

SZ.-NAGY, B., FOIAŞ, C.,

[1] Sur les contractions de l'espace de Hilbert. III. *Acta Sci. Math.* 19 (1958), 26—46

[2] Sur les contractions de l'espace de Hilbert. IV. *Acta Sci. Math.* 21 (1960), 251—259

[3] Sur les contractions de l'espace de Hilbert. V. Translations bilatérales. *Acta Sci. Math.* 23 (1962), 106—129

[4] Sur les contractions de l'espace de Hilbert. VI. Calcul fonctionnel. *Acta Sci. Math.* 23 (1962), 136–167

[5] Dilatation des commutants d'opérateurs. C.R. sér. A, 266 (1968), 493–495

[6] *Analyse Harmonique des Opérateurs de l'espace de Hilbert*. Masson et Cie, Paris, Akadémiai Kiadó, Budapest, 1967 (English transl. North-Holland Publishing Company, Amsterdam–London, Akadémiai-Kiadó, Budapest, 1969)

VAROPOULOS, N. TH.,

[1] On an Inequality of non Neumann and an Application of the Metric Theory of Tensor Products to Operators Theory. *J. Funct. Anal.* 16 (1974), 83–100

WERMER, J.,

[1] On algebras of continuous functions. *Proc. Amer. Math. Soc.*4 (1953), 866–869

[2] Dirichlet algebras. *Duke Math. J.* 27 (1960), 373–382

[3] The space of real parts of a function algebra. *Pacific J. Math.* 13 (1963), 1423–1426

[4] *Banach Algebra and Analytic Functions*. Advances in Math. Academic Press, New York, 1961

WILKEN, D. R.,

[1] Lebesgue measures of parts for *R(X)*. *Proc. Amer. Math. Soc.* 18 (1967), 508–512

WOLD, H.,

[1] *A study in the analysis of stationary time series*. Stockholm, 1938,second ed. 1954

ZERNER, M.,

[1] Fonctionnelles propres des opérateurs sur les espaces nucléaires. *Summer School on Topological Algebra Theory*, Bruges, 1966 (187–232)

YOSIDA, K.,

[1] *Functional Analysis*. Springer, Berlin–Göttingen–Heidelberg, 1965